$\delta^{37}Cl$

The geochemistry of chlorine isotopes

Eggenkamp, Hermanus Gerardus Maria
$\delta^{37}Cl$: The geochemistry of chlorine isotopes
(Originally issued as Ph.D. Thesis, Utrecht University, 1994)
This soft cover edition © 2014
Onderzoek en Beleving (http://www.onderzoek-en-beleving.nl)
ISBN: 978-90-816059-3-9 (Soft cover; CreateSpace)
NUR: 931
Front cover © M.H.C. Vlaanderen

$\delta^{37}Cl$

The geochemistry of chlorine isotopes

De geochemie van chloorisotopen

(met een samenvatting in het nederlands)

PROEFSCHRIFT

TER VERKRIJGING VAN DE GRAAD VAN DOCTOR
AAN DE UNIVERSITEIT UTRECHT
OP GEZAG VAN DE RECTOR MAGNIFICUS,
PROF. DR J.A. VAN GINKEL,
INVOLGE HET BESLUIT VAN HET COLLEGE VAN
DECANEN IN HET OPENBAAR TE VERDEDIGEN OP
MAANDAG 24 JANUARI 1994 DES MIDDAGS TE 14.30 UUR

DOOR

HERMANUS GERARDUS MARIA EGGENKAMP
GEBOREN OP 22 OKTOBER 1963, TE LAREN (NH)

Promotor: Prof. Dr R.D. Schuiling
                   Prof. Dr A.F. Koster van Groos (Chicago, IL, U.S.A.)
Co-promotor: Dr R. Kreulen

Dit proefschrift is mogelijk gemaakt met financiële steun van de stichting Aardwetenschappelijk Onderzoek (Projectnummer 751.355.014), en reisbeurzen van de Nederlandse Organisatie voor Wetenschappelijk Onderzoek, Shell Nederland en de Vakgroep Geochemie.

Aan mijn ouders

Voorwoord

Een studie als hier gepresenteerd kon natuurlijk niet uitgevoerd worden zonder de hulp van velen. Gezien de veelzijdigheid van de onderwerpen ben ik de meeste dank verschuldigd aan allen die belangeloos monsters ter beschikking hebben gesteld, waardoor dit werk verricht kon worden. Jack Middelburg leverde de poriënwaters van Kau Baai, de poriënwaters van het IJsselmeer zijn door Tony Appelo en Hans Beekman ter beschikking gesteld. De formatie water monsters heb ik van Prof. C.H. van der Weijden, Max Coleman (BPX, Sunbury-on-Thames, Engeland) en Jean-Michel Matray (BRGM-IMRG, Orléans, Frankrijk) gekregen. De vulkanische monsters uit Indonesië komen van René Poorter, Jurian Hoogewerff, Johan Varekamp en Rob Kreulen. De minerale waters uit Noord-Portugal zijn bemonsterd door Prof. C.H. van der Weijden, Paul Saager en Ria Wijland, terwijl het oppervlakte van het IJsselmeer samen met Peter van der Poel, Bert Kos en Arjan Bos is bemonsterd. De lithium brines zijn beschikbaar gesteld door Dr I.A. Kunasz van de Foote Mineral Company in Exton, PA (Verenigde Staten), en de diep-zee brine monsters door Gert de Lange. De evaporieten zijn beschikbaar gesteld door Biliton Refractories B.V. te Veendam, in het bijzonder dankzij de heren Ing. H. Lorenzen en Ing. H.P. Rogaar. De Ilímaussaq monsters uit Groenland zijn beschikbaar gesteld door Dr J. Konnerup-Madsen en Dr J. Rose-Hansen van het Institut for Petrologi in Kopenhagen, de testmonsters om de ontsluitingsmethode te testen komen van Geert-Jan de Haas en Bas Dam. De meeste mineraal species zijn afkomstig van het Nationaal Natuurhistorisch Museum in Leiden, met veel dank aan Prof. P.C. Zwaan, Dr C.E.S. Arps en de heer Diederiks, de andere zijn afkomstig uit de collectie van het Mineralogisch-Geologisch Instituut van de Universiteit Utrecht of verkregen van Bas Dam. De carbonatieten tot slot zijn beschikbaar gesteld door Prof. J. Keller van de Albert-Lüdwigs-Universität in Freiburg en Dr M.J. Le Bas van de University of Leicester.

Rob Kreulen, mijn copromotor, wil ik bedanken voor de mogelijkheid die hij mij heeft geboden dit zeer interessante veld van onderzoek in een proefschrift te omschrijven. Hoewel we het niet steeds eens zijn geworden, zou zonder zijn inzet dit proefschrift nooit verschenen zijn.

Heel veel dank ben ik Sven Scholten, mijn kamergenoot gedurende het grootste deel van mijn promotie, verschuldigd. De vele discussies die wij over het onderwerp voerden, en het kritische doorlezen van de manuscripten hebben het geheel sterk verbeterd.

Guus Koster van Groos, één van mijn promotoren, bedank ik voor zijn inzet, om samen met Rob Kreulen een eerste opzet voor de meting van chloorisotopen in Utrecht op te zetten, waar dit proefschrift een rechtstreeks resultaat van is. Ook de uiterst vlotte wijze waarop hij manuscripten corrigeerde is zeer gewaardeerd.

Olaf Schuiling, mijn andere promotor, wil ik bedanken voor het doorlezen van de manuscripten en de correcties die hij erin aanbracht.

Jan Meesterburrie, Anita van Leeuwen en Arnold van Dijk bedank ik voor de gezellige tijd die we op het lab doorbrachten en de δ^{13}C, δ^{18}O en δD analyses die ze voor mij verricht hebben.

Math Kohnen, mijn voorganger op dit project, dank ik voor de voorbereidingen die hij al had getroffen zodat ik niet direct in het diepe hoefde te springen.

De gezelligheid op het instituut, één van de belangrijkste zaken die van invloed zijn op het enthousiasme om aan een operatie als deze te werken, is te danken aan onder andere mijn collega promovendi: Bertil van Os, Sven Scholten, Geert-Jan de Haas, Bas Dam, Giuseppe Frapporti, Else Henneke, Pier de Groot, Simon Vriend, Paulien van Gaans, Marcel Paalman, Jack Middelburg, Jurian Hoogewerff, Pieter Vroon, Patrick van Santvoort, Peter Pruysers en Dick Schipper.

Paul Anten wil ik bedanken voor de ICP-analyses.

De co-auteurs van de verschillende hoofdstukken bedank ik voor de stimulerende discussies die we samen hebben gevoerd om deze hoofdstukken in de gepresenteerde vorm te krijgen. Door hun medewerking had ik in ieder geval niet het idee dat ik de enige was die de geochemie van chloorisotopen interessant vond.

De medewerkers van de bibliotheek bedank ik voor de vlotte wijze waarop zij telkens weer de nodige literatuur boven tafel wisten te krijgen.

Heel veel dank ben ik verschuldigd aan Diane McCartney die als een van de weinigen het proefschrift tenminste drie keer heeft gelezen om het engels erin tot een aanvaardbaar niveau te brengen.

Tot slot wil ik natuurlijk mijn ouders bedanken voor de onvoorwaardelijke steun die ik altijd gedurende zowel mijn studie als mijn promotie van hen heb gekregen.

Samenvatting

INLEIDING

De kleinste deeltjes waaruit elementen bestaan zijn de atomen. Niet alle atomen van een element zijn exact hetzelfde, van sommige elementen zijn er zwaardere en lichtere atomen die men isotopen noemt. Het element chloor heeft twee van deze isotopen, het lichtere chloor-35 (^{35}Cl, 17 protonen en 18 neutronen, ca. 75,77% van alle chloor) en het zwaardere chloor-37 (^{37}Cl, 17 protonen en 20 neutronen, ca. 24,23% van alle chloor). In dit proefschrift wordt het geochemische gedrag van deze chloorisotopen beschreven. De verhouding tussen deze twee isotopen is in de natuur erg constant omdat chloor eigenlijk slechts in één enkele oxydatie toestand voorkomt. De gemeten verschillen zijn erg klein en worden weergegeven als een per mil (‰) afwijking van een standaard. De standaard waarvoor is gekozen is zeewater omdat dit uit een zeer groot en goed gemengd reservoir komt. De gemeten verschillen worden weergegeven als $\delta^{37}Cl$ (delta-chloor-37) volgens de volgende vergelijking:

$$\delta^{37}Cl = \frac{(R_{sample} - R_{standard})}{R_{standard}} * 1000$$

In deze vergelijking is R de verhouding tussen de beide chloor isotopen.

METHODE

De methode die in dit onderzoek (**hoofdstuk 2**) is gebruikt om stabiele chloorisotoop variaties te meten is in eerste instantie ontwikkeld door TAYLOR & GRIMSRUD (1969), en verbeterd door KAUFMANN (1984). De chloor isotopen samenstelling wordt gemeten aan CH_3Cl, dat gevormd is door een reactie van AgCl met CH_3I. De AgCl is neergeslagen door een $AgNO_3$ oplossing aan de chloride houdende oplossing toe te voegen. De massa spectrometer heeft een nauwkeurigheid die beter dan 0,07‰ is voor het meten van de referentie gassen. De nauwkeurigheid van de monstervoorbereidings methode werd gedurende het onderzoek steeds beter, en varieerde van 0,13‰ aan het begin tot 0,06‰ (**hoofdstuk 3**) aan het eind van het onderzoek.

GEDRAG VAN CHLOORISOTOPEN GEDURENDE DIFFUSIE

Omdat het werd verondersteld dat diffusie een belangrijk proces is dat variaties in de chloor isotopen samenstelling kon veroorzaken (zie b.v. DESAULNIERS et al. 1986), is de theoretische fractionatie in een aantal eenvoudige diffusie systemen berekend (**hoofdstuk 4**). Afhankelijk van het type van de chloor bron en de geometrie van de (geologische) formaties zijn de chloor en $\delta^{37}Cl$ profielen duidelijk verschillend. Vier verschillende typen chloor bronnen zijn in acht genomen: 1) diffusie vanuit een constante bron, 2) diffusie vanuit een constante bron met sedimentatie of advectieve poriënwater stroming, 3) diffusie vanuit een eenmalige instroom, en 4) diffusie vanuit een constante instroom. Afhankelijk van het diffusie systeem kunnen grote $\delta^{37}Cl$ variaties worden gevonden. Over het algemeen komen negatieve waarden meer voor dan positieve waarden. Omdat de diffusie coëfficiënt van chloor relatief hoog is, worden grote variaties alleen gevonden in jonge diffusie systemen. Het signaal dempt snel uit in de wat oudere (ca. 10000 jaar) systemen.

Deze resultaten zijn vergeleken met een natuurlijk systeem. Het relatief eenvoudige systeem van Kau Baai, Halmahera, Indonesië is hiervoor gekozen (**hoofdstuk 5**). Gedurende de laatste ijstijd was deze baai een zoetwater meer. Na de koude periode steeg de zeespiegel weer en stroomde zeewater de baai in. Omdat zeewater een hogere dichtheid heeft dan zoet water ging dit naar de bodem van de baai. Vanaf dat moment, zo'n 10000 jaar geleden, begon zout water het sediment in te diffunderen. De sedimentatie snelheid in deze sedimenten is constant, wat bekend is uit zowel ^{14}C metingen als chloor diffusie bepalingen. Deze data konden worden bevestigd en het was mogelijk de diffusie coëfficiënt verhouding van ^{35}Cl en ^{37}Cl te bepalen in een natuurlijk systeem. Het werd gevonden dat de waarde iets boven de door Madorsky & Strauss (1948) en Konstantinov & Bakulin (1965) gemeten waarde lag, nl. 1.0023. Deze data gaven ook aan dat in een systeem met zeer kleine δ^{37}Cl verschillen (in dit systeem maximaal 0,38‰), nog nauwkeurige en significante metingen gedaan kunnen worden.

Het chloor isotopen profiel in een veel complexer systeem, zoals het IJsselmeer is (**hoofdstuk 6**) is ook moeilijker te interpreteren. In dit systeem is niet alleen sedimentatie opgetreden, maar ook erosie, stormen en variaties in de zoutheid van het water. Het is erg moeilijk dit in een diffusie model te berekenen. Twee verschillende modellen zijn voorgesteld. Ten eerste een eenvoudig analytisch model, nl. het diffusie model voor het geval er diffusie op treed vanuit een constante bron, zonder en met sedimentatie. Het tweede model is een numeriek model waarbij zoveel mogelijk rekening kan worden gehouden met de bekende geschiedenis van het IJsselmeer (Beekman 1991, Beekman et al. 1992). In alle modellen zijn de randvoorwaarden zoveel als mogelijk hetzelfde genomen. Grote verschillen zijn gevonden tussen de gemeten en berekende chloor concentraties en δ^{37}Cl waarden. Alleen met behulp van het numerieke model, als rekening werd gehouden met alle historische feiten, was het mogelijk de gemeten waarden te reproduceren met een acceptabele nauwkeurigheid.

De extreme δ^{37}Cl variaties die gevonden zijn in formatie (olieveld) waters (**hoofdstuk 7**), zijn het gevolg van verschillende processen, waarvan diffusie er één is. Andere processen zijn ion-filtratie, menging en oplossing van zoutafzettingen. Omdat diverse processen aan de vorming van deze waters bijdragen is een combinatie met diverse ander metingen noodzakelijk om de processen te begrijpen. Twee verschillende modellen zijn bestudeerd. Het eerste is een model met een negatieve chloor-δ^{37}Cl correlatie, zoals gevonden in formatie water uit het Bekken van Parijs (Frankrijk). Het wordt verondersteld dat deze water monsters het resultaat zijn van menging tussen water dat zoutafzettingen heeft opgelost en water dat mogelijk is intstaan door ion-filtratie (Matray et al. 1993). Een positieve correlatie is gevonden in monsters uit het Forties olieveld (Noordzee) en het Westland (Nederland). In beide systemen is het verondersteld dat water met een laag zoutgehalte, afkomstig van dehydratatie van smectiet in het herkomst gesteente van de olie, is gemengd met zout aquifer water. δ^{37}Cl van het herkomst gesteente water is extreem laag omdat deze gesteenten (schalies) een zeer lage porositeit hebben (Coleman et al. 1993, Eggenkamp & Coleman 1993).

ISOTOOP VARIATIES IN VULKANISCHE WATER EN GAS MONSTERS

Chloor isotoop variaties in vulkanische bron waters en gas condensaten (**hoofdstuk 8**) hebben een opmerkelijke correlatie met de δ^{18}O van deze monsters. Over het algemeen hebben deze monsters een negatieve δ^{37}Cl als de δ^{18}O ook negatief is, en een positieve δ^{37}Cl als de δ^{18}O ook positief is. Monsters met een positieve δ^{18}O liggen ver van de meteorisch water lijn, terwijl de monsters met een negatieve waarde er dicht tegenaan liggen. De monsters met positieve δ^{37}Cl en δ^{18}O waarden zijn waarschijnlijk geothermale waters, die hun chloor gekregen hebben van uitgassend gesteente. Er wordt vermoedt dat HCl dat uit magma ontwijkt is verrijkt aan ^{37}Cl. Het gevolg hiervan is dan dat δ^{37}Cl in het residuaire gesteente laag wordt, zodat δ^{37}Cl in regenwater, dat zijn chloor krijgt van verwerend gesteente ook lager is.

δ^{37}Cl is ook gemeten in gas monsters die verzameld zijn in zogenaamde Giggenbach flessen. Er werd een zeer goede omgekeerde correlatie gevonden tussen de chloor concentratie en extreme $\delta^{37}Cl$ waarden (van -1.56 tot +9.5‰). Deze verschillen hebben naar alle waarschijnlijkheid geen verband met de geologie, maar zijn het gevolg van fractionatie gedurende het monsteren of door analytische effecten.

CHLOOR ISOTOPEN IN ANDERE WATERIGE SYSTEMEN

Isotoop verhoudingen zijn ook bepaald in vier kleinere monster sets (**hoofdstuk 9**). Omdat deze sets klein zijn, zijn de gesignaleerde variaties niet goed begrepen. In mineraal water uit Noord-Portugal is $\delta^{37}Cl$ gemeten in zeven bronnen die zijn gerelateerd aan de Ribama breuk. De verschillen zijn erg klein, maar het lijkt erop dat binnen drie te onderscheiden groepen de $\delta^{37}Cl$ toeneemt van noord naar zuid. Het is niet onmogelijk dat dit een diffusie effect is.

Tien monsters van IJsselmeer oppervlakte water zijn gemeten. Variaties binnen het IJsselmeer zijn niet significant, terwijl twee monster van de aangrenzende Waddenzee iets lagere $\delta^{37}Cl$ waarden hebben. Op dit moment is het nog niet mogelijk een verklaring hiervoor te geven. Variaties gevonden in kleine monsters sets bestaande uit nederlands grond- en kraanwater, lithium-, en diepzee pekels kunnen ook niet verklaard worden.

VARIATIES IN DE ISOTOOP VERHOUDINGEN VAN GESTEENTEN EN MINERALEN

$\delta^{37}Cl$ variaties zijn ook gemeten aan een grote groep gesteente en mineraal monsters. De meest eenvoudige gesteente monsters zijn zout afzettingen. Monsters uit deze afzettingen kunnen eenvoudig in water worden opgelost. $\delta^{37}Cl$ variaties zijn niet erg groot, zelfs niet in de monsters met de hoogste indampingsgraad (**hoofdstuk 10**). De verschillende zout mineralen hebben verschillende fractionatie factoren tussen de oplossing en de neerslag. Voor NaCl is de fractionatie +0,24‰, voor KCl -0,05‰ en voor $MgCl_2.6H_2O$ -0,07‰. Dit verklaart waarom geen extreme $\delta^{37}Cl$ variaties werden gevonden in zout afzettingen, en dat het zelfs mogelijk is dat, nadat een minimum is bereikt, $\delta^{37}Cl$ weer toeneemt in het laatste kristalliserende zout (EGGENKAMP et al. 1993).

Een nieuwe ontsluitingsmethode is ontwikkeld voor gesteente monsters. De hoeveelheid chemicaliën die gebruikt wordt moet, om de chloor concentratie in de uiteindelijke oplossing zo hoog mogelijk te houden, zo klein mogelijk zijn. Gepoederde gesteente monsters zijn opgelost in gesmolten NaOH, en het resultaat is aangezuurd met HNO_3. Om de gevormde silicagel neer te slaan wordt HF toegevoegd. Eventueel nog aanwezige fluoride ionen worden neergeslagen met $Mg(NO_3)_2.6H_2O$ (**hoofdstuk 11**). Vijf gesteente monsters van de Groenlandse Ilímaussaq intrusie zijn op deze wijze gemeten. Hoewel de monsters uit verschillende magmatische fasen kwamen werd geen trend gevonden. Alle monsters, met uitzondering van één, hadden dezelfde $\delta^{37}Cl$ waarde. Vermoedelijk is door de hoge kristallisatie temperatuur van dit gesteente de fractionatie te klein om te worden gemeten.

De $\delta^{37}Cl$ gemeten in diverse mineralen gaf grote verschillen (**hoofdstuk 12**). Hoewel de meeste monsters waarden tussen +1 en -1‰ hadden, hadden enkele monsters sterk afwijkende waarden. Een salmiak monster van de italiaanse vulkaan Etna had een waarde van -4,88‰. Dit werd vermoedelijk veroorzaakt door herhaaldelijk sublimeren van het monster, waardoor het uiteindelijke produkt erg negatief werd. Erg hoge waarden zijn gevonden in zogenaamde oxichloriden. Bijvoorbeeld een atacamiet monster met een $\delta^{37}Cl$ waarde van +5,96‰ is gemeten. Deze hoge waarden, gevormd door hydrothermale processen, kunnen mogelijk gebruikt worden om voor deze mineralen te exploreren (EGGENKAMP 1993). Deze waarden zijn de laagste en de hoogste die ooit in natuurlijke monsters gemeten zijn.

Chloor isotoop verhoudingen zijn ook gemeten in enkele carbonatieten (dat zijn carbonaatrijke stollingsgesteenten). Er zijn significante $\delta^{37}Cl$ variaties gevonden (**hoofdstuk 13**). Omdat van primaire carbonatieten wordt aangenomen dat dit mantel materiaal is, kan mogelijk worden aangenomen dat de $\delta^{37}Cl$ gelijk is aan de $\delta^{37}Cl$ van de mantel. Indien dit waar is zou dit betekenen, omdat $\delta^{37}Cl$ van primaire carbonatieten negatief is, dat de mantel ook negatieve $\delta^{37}Cl$ waarden heeft. Dit zou impliceren dat het chloor dat uit de mantel ontsnapt is verrijkt is aan ^{37}Cl, zodat het residu lagere $\delta^{37}Cl$ waarden heeft dan the oppervlakte reservoir (voornamelijk de oceanen). Dit zou dan overeenkomen met de waarnemingen aan de vulkanische monsters uit Indonesië.

Referenties: zie einde hoofdstuk 1.

Publications

(Parts of) the following chapters have been published as papers in scientific journals:

Chapter 5:
H.G.M. Eggenkamp, J.J. Middelburg & R. Kreulen (1994) **Preferential diffusion of** ^{35}**Cl relative to** ^{37}**Cl in sediments of Kau Bay, Halmahera, Indonesia.** *Chem. Geol. (Isot. Geosc. Section)* **116**:317-325.

Chapter 6:
H.E. Beekman, H.G.M. Egenkamp & C.A.J. Appelo (2011) **An integrated modelling approach to reconstruct complex solute transport mechanisms –** ^{37}**Cl/**^{35}**Cl ratios in sediments from a former brackish lagoon in The Netherlands.** *Appl. Geochem.* **26**:257-268.

Chapter 9:
L. Aires-Barros, J.M. Marques, R.C. Graça, M.J. Matias, C.H. van der Weijden, R. Kreulen & H.G.M. Eggenkamp (1998) **Hot and cold CO$_2$-rich mineral waters in Chaves geothermal area (Northern Portugal).** *Geothermics* **27**:89-107.

Chapter 10:
H.G.M. Eggenkamp, R. Kreulen & A.F. Koster van Groos (1995) **Chlorine stable isotope fractionation in evaporites.** *Geochim. Cosmochim. Acta* **59**:5169-5175.

Chapter 12:
H.G.M. Eggenkamp & R.D. Schuiling (1995) δ^{37}**Cl variations in selected minerals; a possible tool for exploration.** *J. Geochem. Expl.* **55**:249-255.

Chapter 13:
H.G.M. Eggenkamp & A.F. Koster van Groos (1997) **Chlorine stable isotopes in carbonatites: evidence for isotopic heterogeneity in the mantle.** *Chem. Geol.* **140**:137-143.

Parts of chapters 2 and 4 have been republished in:
Hans Eggenkamp (2014) **The geochemistry of stable chlorine and bromine isotopes.** Springer Verlag, Heidelberg. ISBN 978-3-642-28505-9 (Hardcover), 978-3-642-28506-6 (eBook). Available from:
http://www.springer.com/earth+sciences+and+geography/geochemistry/book/978-3-642-28505-9

Contents

Chapter 10: **Chlorine stable isotope fractionation in evaporites**   **111**

H.G.M. Eggenkamp, R. Kreulen, A.F. Koster van Groos

Chapter 11: **Stable chlorine isotopes in rocks. A new method for the extraction of chlorine from rocks. Case study: the Ilímaussaq intrusion, South Greenland**

123

H.G.M. Eggenkamp

CHAPTER 1

Synopsis

INTRODUCTION

In this thesis the geochemistry of the stable isotopes of chlorine will be examined. Chlorine is one of the halogens, the seventh group in the periodic system of elements. This group consists of five elements, fluorine, chlorine, bromine, iodine and astatine.

According to the latest report of the COMMISSION ON ATOMIC WEIGHTS AND ISOTOPIC ABUNDANCES (1991a) the standard atomic weight of chlorine is 35.4527±0.0009.

Isotopes are atoms from the same element, i.e. they have the same amount of protons, but with a different mass. In different isotopes the number of neutrons defines the mass of the isotope. Chlorine, for example has 17 protons and since the stable isotopes have masses 35 and 37 the two stable isotopes have 18 and 20 neutrons respectively. ^{35}Cl is the dominant ion with an abundance of 75.77% (COMMISSION ON ATOMIC WEIGHTS AND ISOTOPIC ABUNDANCES 1991b), and thus the abundance of ^{37}Cl is 24.23%. Beside the two stable isotopes 13 instable isotopes of which two have two isomeric states (^{31}Cl, ß$^+$, $t_{1/2}$=0.15s; ^{32}Cl, ß$^+$, $t_{1/2}$=0.297s; ^{33}Cl, ß$_+$, $t_{1/2}$=2.51s; ^{34m}Cl, ß$^+$ and isomeric transition, $t_{1/2}$=32.2min; ^{34}Cl, ß$^+$, $t_{1/2}$=1.53s; ^{36}Cl, ß$^-$ (98%) and ß$^+$, $t_{1/2}$=3.0*10^{5}y; ^{38m}Cl, I.T., $t_{1/2}$=0.70s; ^{38}Cl, ß$^-$, $t_{1/2}$=37.2m; ^{39}Cl, ß$^-$, $t_{1/2}$=55.7m; ^{40}Cl, ß$^-$, $t_{1/2}$=1.35m; ^{41}Cl, ß$^-$, $t_{1/2}$=34s; ^{42}Cl, ß$^-$, $t_{1/2}$=6.8s; ^{43}Cl, ß$^-$, $t_{1/2}$=3.3s; ^{44}Cl and ^{45}Cl, probably ß$^-$, $t_{1/2}$ unknown; HOLDEN 1990). Because of the relatively long half life ^{36}Cl can be used for geochronological purposes, a few recent papers relating to this subject are FEHN et al. (1992), HERUT et al. (1992), NISHIIZUMI et al. (1991), ZREDA et al. (1991) and PHILLIPS et al. (1991).

Two oxidation states of chlorine are found in terrestrial environments. In virtually all cases it is found in state -I as the chloride ion (Cl$^-$). In the extremely dry Atacama desert in Northern Chile also perchlorates (ClO$_4^-$, oxidation state +VII) are found in amounts up to 0.5% in the nitrate deposits (ERICKSEN 1981).

Since variations in the isotopic composition of the elements generally are very small, the actual isotope ratio is not measured, but the difference between the isotope ratio of the sample and a standard is measured, as this difference can be determined more accurately than the actual isotope ratio. This difference will be expressed as per mil according to:

$$\delta^{37}Cl = \frac{(R_{sample} - R_{standard})}{R_{standard}} * 1000 \tag{1}$$

In this equation is R the isotope ratio between ^{37}Cl and ^{35}Cl. Seawater will be used as standard, since KAUFMANN (1984) and KAUFMANN et al. (1984a) have proved that the chlorine isotopic composition of seawater was constant. They called this standard SMOC (Standard Mean Ocean Chloride). No formal isotopic standard has been prepared so far. In this study all measurements are given as deviations from Madeira 84, a seawater from the Atlantic Ocean taken near Madeira.

GEOCHEMICAL CYCLE OF CHLORINE

Chlorine is one of the so called excess volatiles. That are elements that occur in the surface reservoir in far greater amounts than could be accounted for by weathering processes which produced the sedimentary rocks. Rubey (1951) concluded that the surface reservoir was gradually filled by escape of these volatiles from rocks that rose from deeper parts of the earth. In that time submarine volcanism was not known, but later Anderson (1974, 1975) calculated the Cl and S flux from island arcs and concluded that

Fig 1: Chlorine cycle of the crust (Kaufmann 1984). The reservoir sizes are expressed in kilograms chlorine, the fluxes in kilograms chlorine per year.

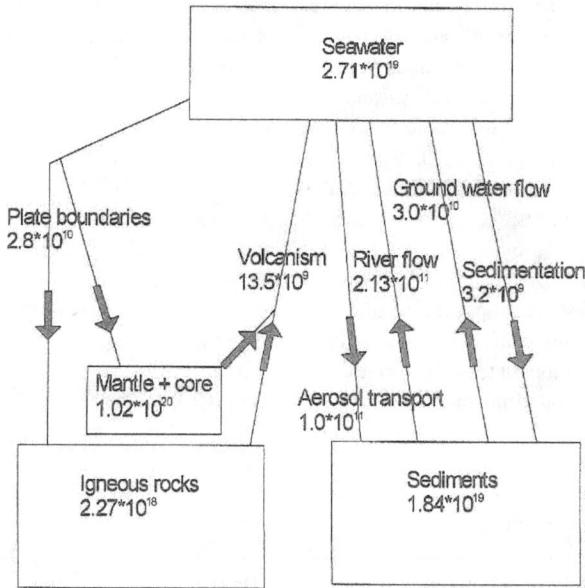

in the earth history 0.3 to 4 times the content could be escaped from these arcs. Schilling et al. (1978) recalculated the flux from the earth's interior to the surface reservoir that nearly all chloride came from hot spots and spreading zones.

The general chlorine cycle is clearly reviewed by Kaufmann (1984), and the mean conclusions for the crustal cycle are summarized in Fig. 1. This review was mainly based on the estimations and assumptions by Schilling et al. (1978). In Fig. 1 the reservoir sizes are replaced by the total amount of chlorine in the reservoirs as calculated by Schilling et al. (1978). Their data can be found in **table 1**, where also the sources of the data are given. All data have very large uncertainties, and it is clear that only the order of magnitude of the fluxes can be estimated.

2

Table 1: *Chlorine in surface reservoirs and fluxes between these reservoirs.*

Amounts of chlorine in reservoirs

Reservoir	Mass (kg)	Conc. Cl⁻ (ppm)
Atmosphere	$5.1*10^{18}$ [1]	$1.6*10^{-3}$ [2]
Seawater	$1.4*10^{21}$ [1]	19353 [3]
Sediments (incl. evaporites)	$2*10^{21}$ [4]	9200 [4]
Continental crust	$1.08*10^{22}$ [5]	210 [6]
Oceanic crust	$4.8*10^{21}$ [5]	48 [7]
Mantle and core	$5.96*10^{24}$ [10]	17 [10]
Earth	$5.98*10^{24}$ [8]	25 [9]

References:
[1]   STACEY (1992)
[2]   RAHN (1976)
[3]   PYTKOWICZ & KESTER (1971)
[4]   GARRELS & MACKENZIE (1972)
[5]   GAST (1972)
[6]   TUREKIAN (1971)
[7]   SCHILLING *et al.* (1978)
[8]   RINGWOOD (1975)
[9]   GANAPATHY & ANDERS (1974)
[10] Calculated from data of lithosphere and complete earth.

HISTORY OF CHLORINE ISOTOPE MEASUREMENTS

It was proven by ASTON (1919) that chlorine consist of two different isotopes with masses 35 and 37. In later years the ratio between the isotopes was determined quit often (e.g. CURIE 1921, GLEDITSCH & SANDAHL 1922, HARKINS & STONE 1925, KALLMAN & LASAREFF 1932, NIER & HANSON 1936, GRAHAM *et al.* 1951, SHIELDS *et al.* 1962). Because in nature the differences in isotope ratios are small, no significant variations were found. Variations in the chlorine isotope ratios were found in chemical experiments (e.g. BARTHOLOMEW *et al.* 1954, KLEMM & LUNDÉN 1955, LUNDÉN & HERZOG 1956, HERZOG & KLEMM 1958, HILL & FRY 1958, HOWALD 1960) and it was found that the diffusion coefficient of ^{35}Cl was about 1.0012 to 1.0022 times that of ^{37}Cl (MADORSKY & STRAUSS 1948, KONSTANTINOV & BAKULIN 1965). After the development of a new mass spectrometer with double ion collectors (NIER *et al.* 1946, NIER 1947, McKINNEY *et al.* 1950, NIER 1955) it was possible to measure the isotope ratio variation with a precision of 1‰. HOERING & PARKER (1961) measured $\delta^{37}Cl$ values of 81 samples. They found no significant variations from the standard, which was first chosen by them to be seawater. Two samples of formation water had large (although not significant) deviations from the standard (-0.7 and -0.8‰), but they were not considered to be different within the precision. They also measured 3 samples of Chilean perchlorate. Although UREY (1947) had predicted that, if hydrogen chloride and perchlorate are in equilibrium the fractionation had to be 92‰, HOERING & PARKER (1961) did not find any difference between the perchlorate and chloride samples and they concluded that the perchlorate was not formed in equilibrium with the chloride in these deposits. MORTON & CATANZARO (1964) measured the chlorine isotope composition from apatites and found no variations within 1‰.

Since the early eighties it is possible to measure chlorine isotope ratio variations with a precision smaller than the natural variations. KAUFMANN (1984) published the first thesis in which measurable variation was proven. The precision of the analyses was better than 0.24‰ and became better in later years. In the same period the first results were presented on several congresses (KAUFMANN et al. 1983, 1984b, KAUFMANN & LONG 1984, CAMPBELL & KAUFMANN 1984) and published (KAUFMANN et al. 1984a). In the years that followed several studies on the geochemistry of the stable isotopes of chlorine were presented by this Arizona group (KAUFMANN et al. 1987, 1988, 1992, 1993 KAUFMANN & ARNÓRSSON 1986, KAUFMANN 1989, DESAUNIERS et al. 1986, EASTOE et al. 1989, EASTOE & GUILBERT 1992, GIFFORD et al. 1985).

A very interesting point of view concerning chlorine isotopes is published by TANAKA & RYE (1991), who suggested to use this ratio as a potential tool for the determination of the origin of chlorine in the atmosphere.

A detailed report with a description of the method to measure $\delta^{37}Cl$ variations as used in Arizona was published recently (LONG et al. 1993)

In the meantime a few studies were presented where the chlorine isotope ratio was measured by Negative Ion Thermal Ionization Mass Spectrometry (compared to gas mass spectrometry in the Arizona and Utrecht measurements). In these studies (VENGOSH et al. 1989 GAUDETTE 1990) the accuracy of the measurements was very poor and the variations extreme. For this reason the samples of these studies are recommended to be remeasured with gas mass spectrometry.

Since 1985 in Utrecht it is tried to measure variations of chlorine isotopes. A research in which chlorine isotope variations are measured with Accelerator Mass Spectrometry is not continued. From 1987 to 1988 the research was done by KOHNEN (1988), and continuad by the present author. In this study chlorine isotope variations are examined in both water and rock samples. One of the main aims was to make an inventory of chlorine isotope variations in different geological systems. For this reason, in this thesis also researches are presented in which very small, or not yet understood variations are found.

THE METHOD

The method used in this research (**chapter 2**) was first developed by TAYLOR & GRIMSRUD (1969) and improved by KAUFMANN (1984). The chlorine isotope composition is measured on CH_3Cl which is formed by reaction of AgCl with CH_3I. The AgCl is produced by precipitation through adding a $AgNO_3$ solution to the chloride containing solution. The mass spectrometer has a precision better than 0.07‰ for the measurement of the reference gasses. The accuracy of the sampling preparation method increased during the research from about 0.13 to 0.06‰ (**chapter 3**), as can be seen from the results of the very often measured standard Madeira seawater.

BEHAVIOUR OF CHLORINE ISOTOPES DURING DIFFUSION

Since it was supposed that diffusion is an important process to cause variations in the isotopic composition (see e.g. DESAULNIERS *et al.* 1986) the theoretical fractionation was calculated in simple diffusion systems (**chapter 4**). Depending on the type of chloride source and the geometry of the formations, the chloride and $\delta^{37}Cl$ profiles are quite different. Four possible kinds of chloride source are considered: 1) diffusion from a constant source, 2) diffusion from a constant source with sedimentation or advective flow of porewater, 3) diffusion from a momentary release, and 4) diffusion from a constant inflow. It is found that depending on the diffusion system large $\delta^{37}Cl$ variation can be found, where negative values will be more important than positive values. Because of the relatively high diffusion coefficient of the chloride ion the largest variations will be found in diffusion systems with young ages. The signal flattens out quickly in older (about 10000 years) systems.

The results were compared with a natural system. For this the simple system of Kau Bay, Halmahera, Indonesia was chosen (**chapter 5**). In the last glacial period this bay was a fresh water lake. After the cold period, when sea level started to rise, salt water flowed over the shallow sill and, because this water has a higher density, it went to the bottom of the bay. From that time, about 10000 years ago, salt water started to diffuse into the fresh water sediment. Beside a constant sedimentation rate was present. This rate is known from both ^{14}C measurements and chloride diffusion data. These data could be confirmed and it was possible to determine the diffusion coefficient ratio in a natural system. It was found to be just above the values measured by MADORSKY & STRAUSS (1948) and KONSTANTINOV & BAKULIN (1965), i.e. 1.0023. These data allowed a test for a system with very small $\delta^{37}Cl$ variations. In the sediment core with the largest differences the range was 0.38‰. Although the differences were extremely small, accurate, significant measurements could be made.

The chlorine isotope profile in a more complex system, such as the Lake IJssel in the Netherlands (**chapter 6**) is more difficult to interpret. In this system a cumulation of sedimentation, erosion, storm surges, and variations in water salinity is found. Calculating this into a diffusion model is very difficult. Two different models are proposed. First a simple analytical model, the diffusion model in the case of a constant source with and without sedimentation. The second model is a numeric model which takes into account the history of the lake and it has possibilities to incorporate the various historical effects that took place in this lake (BEEKMAN 1991, BEEKMAN *et al.* 1992). In the models, the pre-set conditions were taken as much as possible the same. Large differences were found between the measured and calculated chloride concentrations and $\delta^{37}Cl$ values. Only in the numerical program, in which all known historical events were taken into account, was it possible to reproduce the measured values with acceptable precision.

The extreme variations in $\delta^{37}Cl$ that are found in formation waters (**chapter 7**), are a result of different processes, of which diffusion is one. Other processes are ion-filtration, mixing and dissolution of evaporites. So, different processes contribute to form these waters, and a combination of several other measurements is necessary to understand the real processes. Two different models are studied, one with a negative δ^{37}Cl-chloride correlation, and one with a positive δ^{37}Cl-chloride correlation. In the waters from the Paris Basin (France) a negative correlation between δ^{37}Cl and chloride is found. This water is supposed to be the result of mixing between meteoric water that dissolved some evaporite and deep formation water that possibly originated from ion-filtration (MATRAY *et al.* 1993). A positive correlation is found in samples from the Dutch Westland and from the Forties Basin in the North Sea. In both systems it is likely that water, originating from dehydratation of smectite in the source rock with a low salinity, had mixed with aquifer water with high salinity originating from dissolving halite deposits. δ^{37}Cl in the water from the source rock is extreme low because of very effective fractionation processes in these low porosity rocks. Mixing has taken place between the low δ^{37}Cl source rock water and a more concentrated aquifer water (COLEMAN *et al.* 1993, EGGENKAMP & COLEMAN 1993) appears to be the dominant process.

ISOTOPE VARIATIONS IN VOLCANIC WATERS AND GASSES

Chlorine isotopes in volcanic spring waters and condensates (**chapter 8**) seem to have a remarkable correlation with the δ^{18}O of these waters. Generally, δ^{37}Cl is negative when δ^{18}O is also negative, and the reverse is also true. The samples with high (positive) δ^{18}O values lie far from the meteoric water line, while the others are near it. The samples with positive δ^{37}Cl and δ^{18}O values are considered to be geothermal waters. In these waters much of the chloride is supposed to be originated from degassing rocks. It seems that HCl that escapes from magma is enriched in ^{37}Cl. δ^{37}Cl of the residual rock then would be negative, so that δ^{37}Cl in meteoric waters which mainly descent from rock weathering, thus, is negative.

The δ^{37}Cl of gas samples collected in Giggenbach bottles also is measured. A very good inverse correlation between the chloride concentration and the extreme variations in δ^{37}Cl (-1.56 to +9.5‰) is found. These differences are probably not related to geologic effects, but to fractionation during sampling, or perhaps because of analytical effects.

CHLORINE ISOTOPES IN OTHER AQUEOUS SYSTEMS

Isotope ratios are determined in 4 smaller data sets (**chapter 9**). Because these sets are very small, the observed differences are not well understood. In mineral water from Northern Portugal δ^{37}Cl is measured on 7 springs related to the Ribama Fault. Variations are small but it seems that within three recognizable groups the δ^{37}Cl increase from northern to southern samples. It may be that this is an effect of diffusion.

Ten samples were sampled from the surface of the Dutch IJsselmeer. Variations within the IJsselmeer were not significant, whereas the two samples from the adjoining Waddenzee were slightly negative. At this moment, no explanation for the variations can be given. Variations found in small datasets from Dutch ground- and tap-water, lithium brines and deep-sea brines can also not yet be explained yet.

Fig 2: Histograms of all known $\delta^{37}Cl$ values between -2.5 and +2.5‰. In the upper diagram all values presented in this thesis are shown, in the lower all literature data (measured by the Arizona group) are presented.

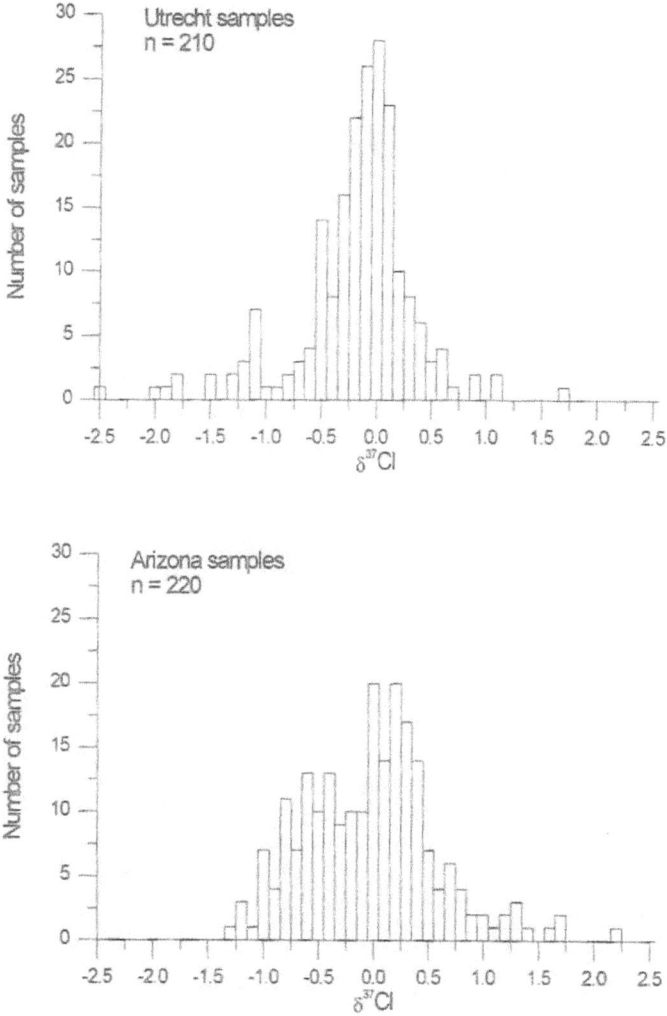

VARIATIONS IN ISOTOPE COMPOSITION OF ROCKS AND MINERALS

A large variety of rock and mineral samples were measured for δ^{37}Cl. The most simple rock samples are evaporites. These samples can be easily dissolved in water. δ^{37}Cl variations are not very large, not even in the samples with the highest degree of evaporation (**chapter 10**). Different salt minerals have different fractionation factors between the solution and the precipitate. For NaCl it is +0.24‰, for KCl it is -0.05‰ and for $MgCl_2.6H_2O$ it is -0.07‰. This explains why no extreme δ^{37}Cl variations were found in evaporite deposits, and that it even is possible that, after a minimum value is reached, the δ^{37}Cl may approach 0‰ in the final crystallizing salts (EGGENKAMP *et al.* 1993).

For rock samples a new dissolution method was developed. The amount of chemicals that need to be added must be as small as possible so that the concentration of chloride in the resulting solution is maximized. For this reason, powdered rock samples are dissolved in molten NaOH, which is acidified with HNO_3. To precipitate the silicic acid HF is added. Remaining fluoride ions are precipitated with $Mg(NO_3).6H_2O$ (**chapter 11**). Five rock samples from the Ilímaussaq intrusion, Greenland were measured. Although the samples came from several magmatic stages no (clear) trend is found, all samples had the same δ^{37}Cl value. It is lilely that at the high crystallization temperature of these rocks fractionation is too small to be measured.

The δ^{37}Cl measured in a set of minerals gave large differences (**chapter 12**). Although most samples have values between -1 and +1‰, a few samples have very different variations; one very negative sample, a sal ammoniac from the Etna, had a δ^{37}Cl of -4.88, the lowest value measured. It seems likely that the lighter isotope (^{35}Cl) sublimes easier than the heavier isotope (^{37}Cl), resulting in a low value for the sublimate. Highly positive values were found in some oxychlorides. For example, an atacamite sample with δ^{37}Cl of +5.96‰ has the highest δ^{37}Cl value measured ever. The isotopic composition for these minerals, formed through hydrothermal processes, is so high that it may be used to explore for these minerals (EGGENKAMP 1993).

Chlorine isotope ratios are also measured in a few carbonatites. Significant differences in chlorine isotope composition are found (**chapter 13**). Since primary carbonatites are thought to present mantle material, the δ^{37}Cl of these rocks may represent δ^{37}Cl of the mantle. If this is the case, which is not known yet, the δ^{37}Cl of the mantle will be negative. This would suggest that chloride that escaped from the mantle is enriched in ^{37}Cl, so that the residue has lower δ^{37}Cl values than the surface reservoir. This would be in agreement with measurements of δ^{37}Cl in volcanic samples from Indonesia.

SUMMARY OF MEASURED δ^{37}CL VALUES

In this study a total of 210 geological samples are measured. From literature data another 220 samples, all measured at the University of Arizona, are extracted (from KAUFMANN *et al.* 1984a, 1987, 1988, 1993, GIFFORD *et al.* 1985, DESAULNIERS *et al.* 1986, KAUFMANN & ARNÓRSSON 1986, EASTOE *et al.* 1989, EASTOE & GUILBERT 1992). Frequency histograms of the two data sets can be found in FIG. 2. Total range is largest in the Utrecht samples (US), from -4.9 to +6.0‰, in the Arizona samples (AS) it is from -1.3 to +2.2‰.

Table 2: *Summary of all measured δ³⁷Cl values.*

	Utrecht	%	Arizona	%	Total	%
$-5.4 \leq \delta^{37}Cl \leq -4.5$	1	0.5	0	-	1	0.2
$-4.4 \leq \delta^{37}Cl \leq -3.5$	2	1.0	0	-	2	0.5
$-3.4 \leq \delta^{37}Cl \leq -2.5$	3	1.4	0	-	3	0.7
$-2.4 \leq \delta^{37}Cl \leq -1.5$	6	2.9	0	-	6	1.4
$-1.4 \leq \delta^{37}Cl \leq -0.5$	37	17.6	56	25.5	85	21.6
$-0.4 \leq \delta^{37}Cl \leq +0.5$	150	71.4	135	61.4	271	66.3
$+0.6 \leq \delta^{37}Cl \leq +1.5$	9	4.3	25	11.4	34	7.9
$+1.6 \leq \delta^{37}Cl \leq +2.5$	1	0.5	4	1.8	5	1.2
$+2.6 \leq \delta^{37}Cl \leq +3.5$	0	-	0	-	0	-
$+3.6 \leq \delta^{37}Cl \leq +4.5$	0	-	0	-	0	-
$+4.6 \leq \delta^{37}Cl \leq +5.5$	0	-	0	-	0	-
$+5.6 \leq \delta^{37}Cl \leq +6.5$	1	0.5	0	-	1	0.2
Total	210	100.0	220	100.0	430	100.0
$-1.4 \leq \delta^{37}Cl \leq +1.5$	196	93.3	216	98.2	412	95.8
$\delta^{37}Cl \geq +0.05$	61	29.0	101	45.9	162	37.7
$-0.04 \leq \delta^{37}Cl \leq +0.04$	28	13.3	20	9.1	48	11.2
$\delta^{37}Cl \leq -0.05$	121	57.6	99	45.0	220	51.2
Average δ³⁷Cl (‰ vs. SMOC)	-0.24		-0.02		-0.13	

These extreme values are very rare. The vast majority of samples have a $\delta^{37}Cl$ between -1.4 and +1.5‰ (over all samples 95.8%), and about two-thirds have a value between -0.4 and +0.5‰. There are striking differences between the two data sets. In the AS many more positive $\delta^{37}Cl$ values are found. In these samples 45.9% has a $\delta^{37}Cl \geq 0.05$‰, in the US only 29.0%. In the AS also the average $\delta^{37}Cl$ value is higher, -0.02‰ relative to -0.24‰ for the US (see also **table 2**). The reason for this difference is probably that the AS include many more samples from geothermal sources. Also, the US data set contains a large set of extremely negative formation waters.

REFERENCES

ANDERSON A.T. (1974) Chlorine, sulfur and water in magmas and oceans. *Geol. Soc. Am. Bull.* **85** 1485-1492

ANDERSON A.T. (1975) Some basaltic and andesitic gases. *Rev. Geophys. Space Phys.* **13** 37-55

ASTON F.W. (1919) The constitution of the elements. *Nature* **104** 393

BARTHOLOMEW R.M., BROWN F. & LOUNSBURY M. (1954) Chlorine isotope effect in reactions of *tert*-butyl chloride. *Canadian J. Chem.* **32** 979-983

BEEKMAN H.E. (1991) *Ion chromatography of fresh- and seawater intrusion. Multicomponent dispersive and diffusive transport in groundwater.* Ph.D. Thesis, Free University, Amsterdam. 198 pp.

BEEKMAN H.E., EGGENKAMP H.G.M., APPELO C.A.J. & KREULEN R. (1992) ^{37}Cl-^{35}Cl transport modelling in accumulation sediments of a former brackish lagoonal environment. *Proc. WRI* **7** 209-212

CAMPBELL D.J. & KAUFMANN R.S. (1984) Fractionation of chlorine isotopes by flow through semi-permeable membranes. *Eos* **65** 882

COLEMAN M., EGGENKAMP H., MATRAY J.M. & PALLANT M. (1993) Origins of oil-field brines by Cl stable isotopes. *Terra Abs.* **5 (1)** 638

COMMISSION ON ATOMIC WEIGHTS AND ISOTOPIC ABUNDANCES (1991a) Atomic weights of the elments 1989. *Pure & Appl. Chem.* **63** 975-990

COMMISSION ON ATOMIC WEIGHTS AND ISOTOPIC ABUNDANCES (1991b) Isotopic compositions of the elements 1989. *Pure & Appl. Chem.* **63** 991-1002

CURIE I. (1921) Sur le poids atomique du chlore dans quelques mineraux. *Compte Rend. Sean.* **172** 1025-1028

DESAULNIERS D.E., KAUFMANN R.S., CHERRY J.A. & BENTLEY H.W. (1986) ^{37}Cl-^{35}Cl variations in a diffusion-controlled groundwater system. *Geochim. Cosmochim. Acta* **50** 1757-1764

EASTOE C.J. & GUILBERT J.M. (1992) Stable chlorine isotopes in hydrothermal systems. *Geochim. Cosmochim. Acta* **56** 4247-4255

EASTOE C.J., GUILBERT J.M. & KAUFMANN R.S. (1989) Preliminary evidence for fractionation of stable chlorine isotopes in ore-forming hydrothermal systems. *Geology* **17** 285-288

EGGENKAMP H.G.M. (1993) Can chlorine stable isotope ratio variations be used in mineral exploration? *Abstract Vol. IGES16.* Beijing. pp. 37-38

EGGENKAMP H.G.M. & COLEMAN M.L. (1993) Extreme $\delta^{37}Cl$ variations in formation water and its possible relation to the migration from source to trap. *Abstract AAPG International Conference and Exhibition.* The Hague.

EGGENKAMP H.G.M., KREULEN R. & KOSTER VAN GROOS A.F. (1993) Fractionation of chlorine isotopes in evaporites. *Terra Abs.* **5 (1)** 650

ERICKSEN G.E. (1981) Geology and origin of the Chilean nitrate deposits. *USGS prof. pap.* **1188** 37 pp.

FEHN U., PETERS E.K., TULLAI-FITZPATRICK S., KUBIK P.W., SHARMA P., TENG R.T.D., GOVE H.E & ELMORE D. (1992) ^{129}I and ^{36}Cl concentrations in waters of the western Clear Lake area, California; residence times and source ages of hydrothermal fluids. *Geochim. Cosmochim. Acta* **56** 2069-2079

GANAPATHY R. & ANDERS E. (1975) Bulk composition of the moon and earth estimated from meteorites. *Geochim. Cosmochim. Acta, Suppl.* **5** 1181-1206

GARRELS R.M. & MACKENZIE F.T. (1972) A quantitative model for the sedimentary rock cycle. *Mar. Chem.* **1** 27-41

GAUDETTE H.E. (1990) Chlorine and boron isotopic analyses of Antarctic ice and snow: indicators of marine and volcanic atmospheric inputs. *Geol. Soc. Amer. Ann. Meeting* **1990** 173

GAST P.W. (1972) The chemical composition of the earth, the moon and chondritic meteorites. *in Nature of the solic Earth.* (Edt. E.C. ROBERTSON) McGraw-Hill, New York. 19-40

GIFFORD S., BENTLEY H. & GRAHAM D.L. (1985) Chlorine isotopes as environmental tracers in Columbia River basalt groundwaters. *Proc. 17th Int. Congr. I.A.H.* Vol. **17** Hydrogeology of rocks of low permeability. 417-429

GLEDITSCH E. & SAMDAHL B. (1922) Radioactivité sur le poids atomique de chlore dans un mineraux ancien, l'apatede Balme. *Compte Rend. Sean.* **174** 746-748

GRAHAM R.P., MACNAMARA J., CROCKER I.H. & MACFARLENE R.B. (1951) The isotopic constitution of germanium. *Canadian J. Chem.* **29** 89-102

HARKINS W.D. & STONE S.B. (1926) The isotopic composition and atomic weight of chlorine from meteorites and from minerals of non-marine origin. *J. Amer. Chem. Soc.* **48** 938-949

HERUT B., STARINSKY A., KATZ A., PAUL M. BOARETTO E. & BERKOVITZ D. (1992) ^{36}Cl in chloride-rich rainwater, Israel. *Earth Planet. Sci. Lett.* **109** 179-183

HERZOG W. & KLEMM A. (1958) Die Temperaturabhängigkeit des Isotopie-Effektes bei der elektrolytischen Wanderungen der Chlorionen in geschmolzenem Thallium(I)-chlorid. *Z. Naturforschg.* **13a** 7-16

HILL J.W. & FRY A. (1962) Chlorine isotope effects in the reactions of benzyl and substituted benzyl chlorides with various nucleophiles. *J. Amer. Chem. Soc.* **84** 2763-2769

HOERING T.C. & PARKER P.L. (1961) The geochemistry of the stable isotopes of chlorine. *Geochim. Cosmochim. Acta* **23** 186-199

HOLDEN N.E. (1990) Table of the isotopes (Revised 1990). in: *Handbook of chemistry and physics.* 71st Edition, Edt. D.R. LIDE, CRC Press.

HOWALD R.A. (1960) Ion pairs. I. Isotope effects shown by chloride solutions in glacial acetic acid. *J. Amer. Chem. Soc.* **82** 20-24

KAUFMANN R.S. (1984) *Chlorine in ground water: stable isotope distribution.* Ph.D Thesis, University of Arizona, Tucson. 137 pp.

KAUFMANN R.S. (1989) Equilibrium exchange models for chlorine stable isotope fractionation in high temperature environments. *Proc. WRI* **6** 365-368

KAUFMANN R.S. & ARNÓRSSON S. (1986) ^{37}Cl/^{35}Cl ratios in Icelandic geothermal waters. *Proc. WRI* **5** 325-327

KAUFMANN R.S. & LONG A. (1984) Stable chlorine ratio as a ground-water tracer. *Eos* **65** 886

KAUFMANN R., LONG A., BENTLEY H. & DAVIS S. (1983) Application of chloride stable isotope analyses to hydrogeology. *Proc. 1983 meet. Amer. Water Res. Ass., Ariz. & Ariz.-Nev. Acad. Sci., Hydr. sect. Hydr. Water Res. Ariz. Southwest* **13** 85-90

KAUFMANN R., LONG A., BENTLEY H. & DAVIS S. (1984a) Natural chlorine isotope variations. *Nature* **309** 338-340

KAUFMANN R.S., LONG A., DAVIS S. & BENTLEY H. (1984b) Natural variations of chlorine stable isotopes. *Abstr. Prog. Geol. Soc. Amer.* **16** 556

KAUFMANN R., FRAPE S., FRITZ P. & BENTLEY H. (1987) Chlorine stable isotope composition of Canadian Shield brines. in: *Saline water and gases in crystalline rocks.* Edt. FRITZ P. & FRAPE S.K. *Geol. Ass. Can. Spec. Pap.* **33** 89-93

KAUFMANN R.S., LONG A. & CAMPBELL D.J. (1988) Chlorine isotope distribution in formation waters, Texas and Louisiana. *AAPG Bull.* **72** 839-844

KAUFMANN R.S., FRAPE S.K., MCNUTT R. & EASTOE C. (1992) Chlorine stable isotope distribution of Michigan Basin and Canadian Shield formation waters. *Proc. WRI* **7** 943-946

KAUFMANN R.S., FRAPE S.K., MCNUTT R. & EASTOE C. (1993) Chlorine stable isotope distribution of Michigan basin formation waters. *Appl. Geoch.* **8** 403-407

KLEMM A. & LUNDÉN A. (1955) Isotopenanreicherung beim Chlor durch electrolytische Überfürung in geschmolzenem Bleichlorid. *Z. Naturforschg.* **10a** 282-284

KOHNEN M.E.L. (1988) *Stabiele chloorisotopen onderzoek.* Internal report, University of Utrecht. 17 pp.

KONSTANTINOV B.P. & BAKULIN E.A. (1965) Separation of chloride isotopes in aqueous solutions of lithium chloride, sodium chloride, and hydrochloric acid. *Russ. J. Phys. Chem.* **39** 315-318

LONG A., EASTOE C.J., KAUFMANN R.S., MARTIN J.G., WIRT L. & FINLEY (1993) High-precision measurement of chlorine stable isotope ratios. *Geochim. Cosmochim. Acta* **57** 2907-2912

LUNDÉN A. & HERZOG W. (1956) Isotopenanreicherung bei Chlor durch electrolytische Überführung in geschmolzenem Zinkchlorid. *Z. Naturforschg.* **11a** 520

MADORSKY S.L. & STRAUSS S. (1947) Concentration of isotopes of chlorine by the counter-current electromigration method. *J. Res. Nat. Bur. Stand.* **38** 185-189

MATRAY J.M., COLEMAN M.L. & EGGENKAMP H.G.M. (1993) Origin of the Keuper formation waters in the Paris Basin. in *Geofluids '93 Extended Abstracts* Edt. J. PARNELL *et al.* 319-322

MCKINNEY C.R., MCCREA J.M., EPSTEIN S., ALLEN H.A. & UREY H.C. (1950) Improvements in mass spectrometers for the measurement of small differences in isotope abundance ratios. *Rev. Sci. Inst.* **21** 724-730

MORTON R.D. & CATANZARO E.J. (1964) Stable chlorine isotope abundances in apatites from Ødegårdens Verk, Norway. *Norsk Geol. Tiddsk.* **44** 307-313

NIER A.O. (1947) A mass spectrometer for isotope and gas analysis. *Rev. Sci. Inst.* **18** 398-411

NIER A.O. (1955) Determination of isotopic masses and abundances by mass spectrometry. *Science* **121** 737-744

NIER A.O. & HANSON E.E. (1936) A mass-spectrographic analysis of the ions produced in HCl under electron impact. *Phys. Rev.* **50** 722-726

NIER A.O., NEY E.P. & INGHRAM M.G. (1946) A null method for the comparison of two ion currents in a mass spectrometer. *Phys. Rev.* **70** 116-117

NISHIIZUMI K., ARNOLD J.R., KLEIN J., FINK D., MIDDLETON R. KUBIK P.W. SHARMA P., ELMORE D. & REEDY R.C. (1991) Exposure histories of lunal meteorites; ALHA81005, MAC88104, MAC88105, and Y791197. *Geochim. Cosmochim. Acta* **55** 3149-3155

PHILLIPS F.M., ZREDA M.G., SMITH S.S., ELMORE D., KUBIK P.W. DORN R.I. and RODDY D.J. (1991) Age and geomorphic history of Meteor Crater, Arizona, from cosmogenic ^{36}Cl and ^{14}C in rock varnish. *Geochim. Cosmochim. Acta* **55** 2695-2698

PYTKOWICZ R.M. & KESTER D.R. (1971) Physical chemistry of sea water. *Ocanogr. Mar. Biol. A. Rev.* **9** 11-60

RAHN K.A. (1976) *Tech. Rep.* Univ. Rhode Island.

RINGWOOD A.E. (1975) *Composition and petrology of the Earth's mantle.* McGraw-Hill, New York. 618 pp.

RUBEY W.W. (1951) Geologic history of seawater: an attempt to state the problem. *Geol. Soc. Am. Bull.* **62** 1111-1148

SCHILLING J.-G., UNNI C.K. & BENDER M.L. (1978) Origin of chlorine and bromine in the oceans. *Nature* **273** 631-636

SHIELDS W.R, MURPHY T.J., GARNER E.L. & DIBELER V.H. (1962) Absolute isotopic abundance ratios and the isotopic weight of chlorine. *J. Amer. Chem. Soc.* **84** 1519-1522

STACY F.D. (1992) *Physics of the Earth.* 3rd edition. Whiley Interscience, New York. 513 pp.

TANAKA N. & RYE D.M. (1991) Chlorine in the stratosphere. *Nature* **353** 707

TAYLOR J.W. & GRIMSRUD E.P. (1969) Chlorine isotopic ratios by negative ion mass spectrometery. *Anal. Chem.* **41** 805-810

TUREKIAN K.K. (1971) *Encyclopedia of Science and Technology.* 2nd Ed., McGraw-Hill, New York. 627-630

UREY H. (1947) The thermodynamic properties of isotopic substances. *J. Chem. Soc.* **69** 562-581

VENGOSH A., CHIVAS A.R. & MCCULLOCH M.T. (1989) Direct determination of boron and chlorine isotopic compositions in geological materials by negative thermal-ionization mass spectrometry. *Chem. Geol. (Isot. Geosci. Sect.)* **79** 333-343

VON KALLMAN H. & LASAREFF W. (1932) Über die Isotopenuntersuchungen (Sauerstoff, Neon und Chlor). *Z. F. Phys.* **80** 237-241

ZREDA M.G., PHILLIPS F.M., ELMORE D., KUBIK P.W. SHARMA P. & DORN R.I. (1991) Cosmogenic chlorine-36 production rates in terrestrial rocks. *Earth Planet. Sci. Lett.* **105** 94-109

CHAPTER 2

Analytical Procedures for δ^{37}Cl Measurements

H.G.M. Eggenkamp[1], M.E.L. Kohnen[2] and R. Kreulen[1]

ABSTRACT-- The isotope composition of chlorine is measured on chloromethane (CH$_3$Cl) gas. This gas is produced from dissolved chloride as follows. The chloride is precipitated with silver nitrate as silver chloride (AgCl). Next, it is reacted with iodomethane (CH$_3$I) to form chloromethane, which is subsequently purified by gas chromatography. The chlorine isotopic content of the purified gas is measured on the mass spectrometer using the differences in the masses 50 and 52.

INTRODUCTION

In the described method δ^{37}Cl was determined by converting chlorine in the samples to chloromethane (CH$_3$Cl), after which masses 52 and 50 were measured in the mass spectrometer. In early chlorine isotope studies gasses such as HCl (HOERING & PARKER 1961), Cl$_2$ (BARTHOLOMEW *et al.* 1954), or COCl$_2$ (ASTON 1941) were used. These gasses, however are in sufficiently inert and, therefore, cause large memory effects. Two different methods can be used to produce chloromethane from chloride. The first method is the reaction of ammonium chloride (NH$_4$Cl) with dimethyl sulphate ([CH$_3$]$_2$SO$_4$):

$$[CH_3]SO_4 + 2NH_4Cl \Leftrightarrow 2CH_3Cl + [NH_4]SO_4 \qquad (1)$$

A major disadvantage of this method is that the CH$_3$Cl yield is only about 35% (OWEN & SCHAEFFER 1955) which produces unwanted isotope effects and, therefore, inaccurate measurements. The second method uses the reaction of silver chloride (AgCl) with iodomethane (CH$_3$I):

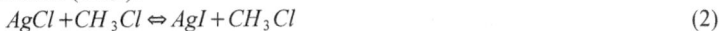

$$AgCl + CH_3Cl \Leftrightarrow AgI + CH_3Cl \qquad (2)$$

LANGVAD (1954) developed a procedure to obtain high chloromethane yields, which, after modification by HILL & FRY (1962) has a yield of 98 %. TAYLOR & GRIMSRUD (1969) found that this method still produced isotope fractionation and made furter changes which solved the problem. Additional improvements were subsequently made by KAUFMANN (1984) and by ourselves. TAYLOR & GRIMSRUD (1969) used negative ion mass spectrometry, whereas KAUFMANN (1984) and we use positive ion mass spectrometry.

In our procedure, we used the following three steps to produce unfractionated chloromethane of sufficient purity for isotope measurement:
1) precipitation of silver chloride
2) reaction of silver chloride with iodomethane
3) separation by gas chromatography.

1Department of Geochemistry, Utrecht University, P.O.Box 80.021, 3508 TA Utrecht, The Netherlands
2Koninklijke/Shell Exploratie en Productie Laboratorium, Volmerlaan 6, 2288 GD Rijswijk (ZH), The Netherlands

PRECIPITATION OF SILVER CHLORIDE

The procedure to prepare the silver chloride depends slightly on the amount of chloride in the solution. The method aims at precipitating silver chloride from solutions of fixed Cl⁻ amount, fixed ionic strength and fixed pH.

KOHNEN (1988) found that the best results are obtained when the amount of silver chloride formed is about $1*10^{-4}$ mole (or 14.3 mg AgCl, corresponding to 3.5 mg Cl⁻). This was confirmed by our later studies (see chapter 3). Therefore the amount of chloride solution needed is:

$$\frac{3000}{ppm\,chloride} = ml\ necessary \qquad (3)$$

If the amount of solution is less than 10 ml, the following standard procedure is used: 4 ml of a 1 M KNO_3 solution and 2 ml of a Na_2HPO_4-citric acid buffer solution are added to the chloride solution.

The purpose of the KNO_3 solution is to reach a high ionic strength. TAYLOR & GRIMSRUD (1969) found that using a less than 0.4 M KNO_3 solution gives too low chloromethane yields; for instance a 0.2 M KNO_3 solution gives only 45% yield. The reason for this effect probably is that smaller crystals form at a high ionic strength. These small crystals can react completely whereas larger crystals form a coating of silver iodide that prevents the inner part of the crystals from reacting. Incomplete reaction inevitably leads to fractionation; TAYLOR & GRIMSRUD (1969) found a fractionation of +0.43‰ due to this effect.

The Na_2HPO_4-citric acid buffer solution is used to buffer pH at 2.2. This is necessary to remove small amounts of sulphide which otherwise precipitate as Ag_2S (KAUFMANN 1984), and also to prevent precipitation of other silver salts such as phosphate and carbonate (VOGEL 1951). We used a buffer solution after McILVAINE (1921) which contains 0.71 gr. (0.004 mole) $Na_2HPO_4.2H_2O$ and 20.6 gr. (0.098 mole) $HOC(CH_2CO_2H)_2CO_2H.H_2O$ (citric acid) per liter.

After adding the KNO_3 solution and Na_2HPO_4-citric acid buffer, the mixture is placed on a boiling ring and heated to about 80 °C. Then 1 ml of a 0.2 M $AgNO_3$ solution is added and AgCl starts precipitating instantaneously. The solution is not stirred because the newly formed AgCl will clot and it is difficult to remove it from the stirrer. The suspension then is filtered over a Whatman® glass fibre filter, type GF/F with a retention of 0.7 μm and a standardized filter speed of 6 ml/sec. During filtration the suspension is rinsed with a dilute nitric acid solution (1 ml concentrated HNO_3 in 500 ml water). When the silver chloride precipitate is rinsed with pure water, it occasionally will become colloidal and pass through the filter. Therefore the rinsing solution must contain an electrolyte; nitric acid is chosen because it has no reaction on the precipitate and leaves no residue upon drying (VOGEL 1951).

After filtration, the filter with the precipitate is dried at 80 °C overnight. Care must be taken to protect the silver chloride against light. Silver chloride decomposes under the influence of light according to the reaction:

$$2AgCl \Leftrightarrow 2Ag + Cl_2 \uparrow \qquad (4)$$

Table 1: *Theoretical ion activity products of chloride containing solutions and the added silver nitrate. For the calculations, it is assumed that the total amount of solution is 3000 devided by the chloride concentration (in ml).*

mol AgNO$_3$ added	$2*10^{-4}$	$4*10^{-4}$	$1*10^{-3}$	$2*10^{-3}$	$4*10^{-3}$	$1*10^{-2}$	$2*10^{-2}$
Cl$^-$ conc. in sample (ppm)							
10000	$1.88*10^{-4}$	$3.76*10^{-4}$	$9.40*10^{-4}$	$1.88*10^{-3}$	$3.76*10^{-3}$	$9.40*10^{-3}$	$1.88*10^{-2}$
5000	$4.70*10^{-5}$	$9.40*10^{-5}$	$2.35*10^{-4}$	$4.70*10^{-4}$	$9.40*10^{-4}$	$2.35*10^{-3}$	$4.70*10^{-3}$
2000	$7.52*10^{-6}$	$1.50*10^{-5}$	$3.76*10^{-5}$	$7.52*10^{-5}$	$1.50*10^{-4}$	$3.76*10^{-4}$	$7.52*10^{-4}$
1000	$1.88*10^{-6}$	$3.76*10^{-6}$	$9.40*10^{-6}$	$1.88*10^{-5}$	$3.76*10^{-5}$	$9.40*10^{-5}$	$1.88*10^{-4}$
500	$4.70*10^{-7}$	$9.40*10^{-7}$	$2.35*10^{-6}$	$4.70*10^{-6}$	$9.40*10^{-6}$	$2.35*10^{-5}$	$4.70*10^{-5}$
200	$7.52*10^{-8}$	$1.50*10^{-7}$	$3.76*10^{-7}$	$7.52*10^{-7}$	$1.50*10^{-6}$	$3.76*10^{-6}$	$7.52*10^{-6}$
100	$1.88*10^{-8}$	$3.76*10^{-8}$	$9.40*10^{-8}$	$1.88*10^{-7}$	$3.76*10^{-7}$	$9.40*10^{-7}$	$1.88*10^{-6}$
50	$4.70*10^{-9}$	$9.40*10^{-9}$	$2.35*10^{-8}$	$4.70*10^{-8}$	$9.40*10^{-8}$	$2.35*10^{-7}$	$4.70*10^{-7}$
20	$7.52*10^{-10}$	$1.50*10^{-9}$	$3.76*10^{-9}$	$7.52*10^{-9}$	$1.50*10^{-8}$	$3.76*10^{-8}$	$7.52*10^{-8}$
10	$1.88*10^{-10}$	$3.76*10^{-10}$	$9.40*10^{-10}$	$1.88*10^{-9}$	$3.76*10^{-9}$	$9.40*10^{-9}$	$1.88*10^{-8}$
5	$4.70*10^{-11}$	$9.40*10^{-11}$	$2.35*10^{-10}$	$4.70*10^{-10}$	$9.40*10^{-10}$	$2.35*10^{-9}$	$4.70*10^{-9}$
2	$7.52*10^{-12}$	$1.50*10^{-11}$	$3.76*10^{-11}$	$7.52*10^{-11}$	$1.50*10^{-10}$	$3.76*10^{-10}$	$7.52*10^{-10}$
1	$1.88*10^{-12}$	$3.76*10^{-12}$	$9.40*10^{-12}$	$1.88*10^{-11}$	$3.76*10^{-11}$	$9.40*10^{-11}$	$1.88*10^{-10}$

Therefore the filter with silver chloride is covered with aluminium foil. The aluminium foil must not be in contact with the silver chloride otherwise the aluminium will reduce the silver chloride:

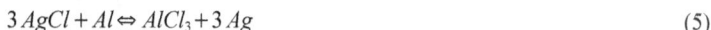

$$3\,AgCl + Al \Leftrightarrow AlCl_3 + 3\,Ag \tag{5}$$

which may cause isotope fractionation. The filter is weighed before precipitation and after drying, so that the amount of silver chloride is known.

Samples with a chloride content below 300 ppm are treated in a slightly different way, because more sample solution is needed. For these samples, KNO$_3$ and the pH buffer are added as dry chemicals, otherwise the amount of solution would become too large. Per 100 ml of sample solution 6.00 g (0.06 mole) KNO$_3$, 2.06 g (0.0098 mole) citric acid and 0.07 g (0.0004 mole) Na$_2$HPO$_4$.2H$_2$O are added.

Samples with a very high pH and dissolved ions (for example in Giggenbach bottles) are diluted in a 1:100 ratio. This solution is treated as stated above.

The chloride solutions that result from dissolving (silicate) rock samples have an ionic strength that is high enough, and a pH that is low enough. Therefore, the AgNO$_3$ can be added directly to these solutions.

-Minimum chloride concentration as a consequence of silver chloride solubility

The methods described above depend on the solubility of AgCl. Although AgCl is known as an "insoluble compound", it is nevertheless soluble to some extent. Its solubility product ranges from $2.1*10^{-11}$ at 4.7 °C to $2.15*10^{-8}$ at 100 °C. At the temperature the precipitation is normally made (about 80 °C) the solubility product is about $1*10^{-8}$. If the product of Ag$^+$ and Cl$^-$ ions is lower than this value no precipitation will take place. **Table 1** shows the ion products calculated for various chloride concentrations in the sample solution, in combination with the added amount of AgNO$_3$.

Table 2: *Theoretical amount of AgCl that will precipitate from chloride containing solutions. For the calculations it is assumed that the total amount of solution is 3000 devided by the chloride concentration (in ml). The solubility product of AgCl is assumed to be* $1*10^{-8}$

mol AgNO$_3$ added	$2*10^{-4}$	$4*10^{-4}$	$1*10^{-3}$	$2*10^{-3}$	$4*10^{-3}$	$1*10^{-2}$	$2*10^{-2}$
Cl$^-$ conc. in sample (ppm)							
10000	100	100	100	100	100	100	100
5000	100	100	100	100	100	100	100
2000	100	100	100	100	100	100	100
1000	99	100	100	100	100	100	100
500	98	99	100	100	100	100	100
200	87	93	97	99	99	100	100
100	47	73	89	95	97	99	99
50	0	0	57	89	89	96	98
20	0	0	0	34	34	73	87
10	0	0	0	0	0	0	47
5	0	0	0	0	0	0	0
2	0	0	0	0	0	0	0
1	0	0	0	0	0	0	0

The proportion of chloride that is precipitated can be found in **table 2**. Solutions containing more than 500 ppm Cl$^-$ are no problem; almost all the Cl$^-$ will precipitate, with only 1 ml 0.2 M AgNO₃. Problems arise when chloride concentrations are lower. For concentrations down to about 50 ppm the problem can be overcome by adding more AgNO₃. For lower concentrations the amount of silver to add would be too high; in these cases it will be necessary to preconcentrate the sample.

PREPARATION OF REACTION TUBES

The reaction of AgCl to CH₃Cl takes place in evacuated Pyrex tubes sealed at both ends; the tubes are 8-10 cm long and have an inner diameter of 8 mm and an outer diameter of 12 mm. The filter with AgCl is loaded in a tube sealed at one end, a capillary drawn at the other end, and the tube is evacuated to a pressure less than $2*10^{-1}$ mbar. The tube is then filled with nitrogen gas and sealed with a rubber stopper to prevent air coming in. In a fume-hood, 200 μl (3.21 mmole) of iodomethane (CH₃I) is added. Back on the vacuum line the CH₃I is frozen on the AgCl with liquid nitrogen and the tube is pumped to less than $1*10^{-1}$ mbar. The tube is then sealed at the site of the capillary. The sealed tube is placed in an oven at a temperature of 70 to 80 °C for 48 hours so that the following reaction takes place:

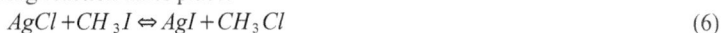

$$AgCl + CH_3I \Leftrightarrow AgI + CH_3Cl \qquad (6)$$

This is an equilibrium reaction so the CH₃I must be added well in excess to get good CH₃Cl yields. If the reaction temperature is too high the CH₃I will partly decompose:

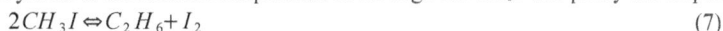

$$2CH_3I \Leftrightarrow C_2H_6 + I_2 \qquad (7)$$

see EASTOE *et al.* (1989). When CH₃I decomposes, the colourless liquid will become yellow to brown. Samples that have been even slightly overheated give much less accurate $\delta^{37}Cl$ values. Decomposition of CH₃I can also been detected in a background scan that is routinely made after the isotope measurement. Samples that have been overheated show an increased background of masses 29, 45 and 46 (FIG. 1)

SEPARATION OF CH₃Cl FROM EXCESS CH₃I

CH₃Cl and CH₃I are separated by gas chromatography on a 75 cm long, ¼" OD SS column, filled with Porapak-Q 80-100 mesh. Because the column is easily overloaded with the large amount of excess CH₃I, the gasses are separated twice so that the remaining CH₃Cl is very pure. The carrier gas is helium, at a pressure of 3 atm and a gas flow of about 100 ml.min⁻¹. The column temperature is 140 °C. A schematic drawing of the setup is shown in FIG. 2.

The procedure is as follows: the gc is in a back-flushing position, in order to minimize CH₃I contamination of the column and detector. The Pyrex reaction tube is scratched by a glass cutting knife, and placed in the crushing tube. The crusing tube is evacuated (FIG. 2A) and liquid nitrogen is placed around the first cold-trap. Valves 1 and 2 are closed and the reaction tube is broken. At the same time valve 3 is turned to position B and valve X is opened. After 30 seconds valves 1 and 2 are turned to position B and valve X is closed (FIG. 2B). After 4½ minutes the liquid nitrogen around the first cold-trap is replaced by warm water and the liquid nitrogen is now placed around the second

FIG. 1. *Background scans of a sample which was not overheated (analysis 7354) and of a sample that was overheated (analysis 7352, note the increased background of peaks 29,45 and 46)*

Background scan for Analysis No. 7359

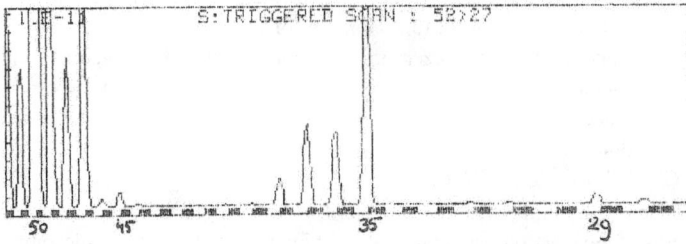

Background scan for Analysis No. 7341

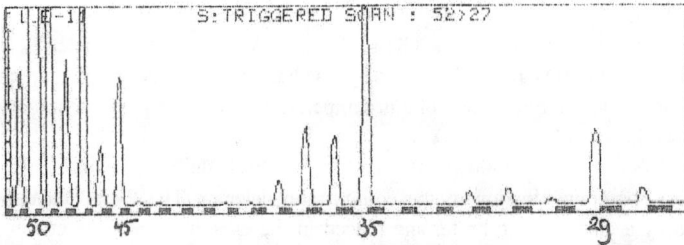

cold-trap. The recorder is started and the CH_3Cl peak will be detected after about 2 minutes. If the recorder is back on the basis line, before the CH_3I peak is detected (after

20

FIG. 2. Schematic drawing of the gas chromatograph.

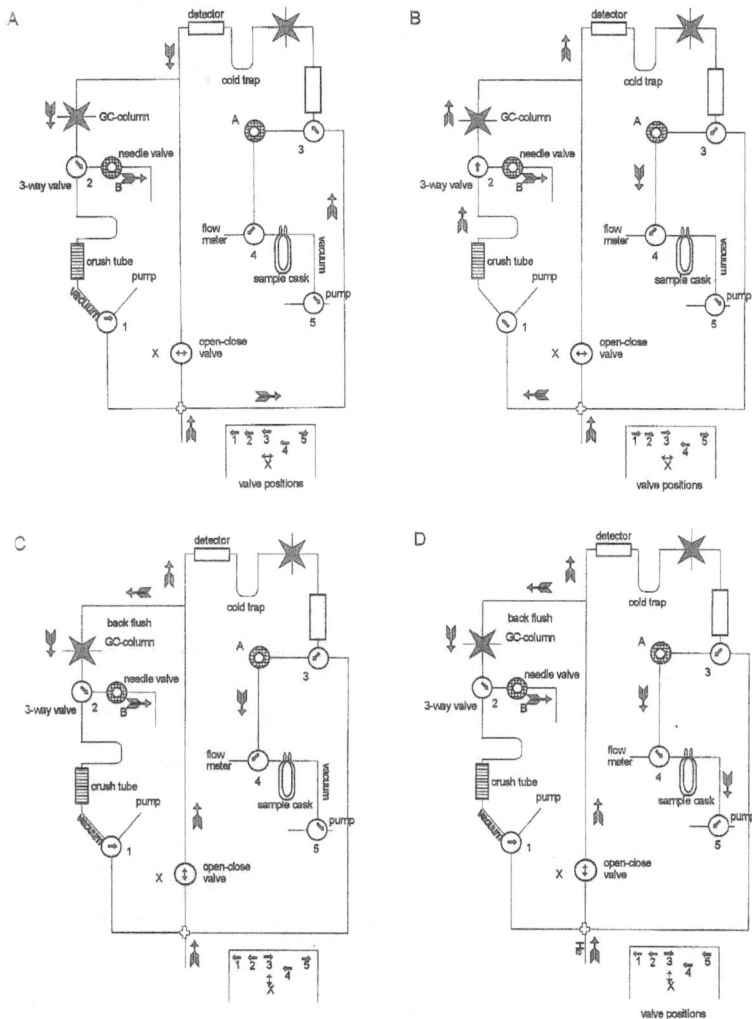

about 4 minutes) valves 1 and 2 are turned to position A and valve X is opened and nitrogen around the second cold-trap is replaced by warm water (FIG. 2C). Liquid nitrogen is now placed around the sample cask. Valve 5 is closed and just before the expected arrival of CH_3Cl valve 4 is turned to position B. As the pressure in the sample bottle becomes higher than 1 atm, valve 5 is turned to position A so that the over-pressure of helium can flow away (FIG. 2D). The CH_3Cl is trapped in the sample-cask. When the

21

whole peak is trapped valve 5 is closed and the other valves must go back in the starting position (Fig. 2A). Helium is pumped out of the sample-cask and the CH_3Cl yield is determined by measuring the pressure. The column is now backflushing so that remaining CH_3I is removed. The broken reaction tube can be replaced by another one.

MEASURING THE SAMPLES WITH THE MASS SPECTROMETER

All samples are measured on the VG SIRA 24 EM mass spectrometer of the department of geochemistry of the university of Utrecht. $\delta^{37}Cl$ is determined from the beams of mass 52 ($CH_3{}^{37}Cl^+$) in collector 3 and mass 50 ($CH_3{}^{35}Cl^+$) in collector 1. The isotope ratio of chlorine is much higher than for the light elements for which the mass spectrometer was built. Thus, beam 52 will be off scale at small working pressures. Working with very low pressures gives isotope fractionation in the inlet system. For this reason the ion source is made less sensitive. This is done by reducing the trap current to $100\mu A$. In this case the minor beam is brought to a value smaller than $10^{-10}A$ while still maintaining sufficient gas pressure in the inlet system. At these conditions, the results are highly reprodicible, as will be shown in the next chapter.

REFERENCES

Aston F.W. (1941) *Mass spectra and isotopes.* Longmans, Green and Co. New York.

Bartholomew R.M., Brown F. & Lounsbury M. (1954) Chlorine isotope effect in reactions of *tert*-butyl chloride. *Canadian J. Chem.* **32** 979-983

Eastoe C.J., Guilbert J.M. & Kaufmann R.S. (1989) Preliminary evidence for fractionation of stable chlorine isotopes in ore forming hydrothermal systems. *Geology* **17** 285-288

Hill J.W. & Fry A. (1962) Chlorine isotope effects in the reactions of benzyl and substituted benzyl chlorides vith various nucleophiles. *J. Amer. Chem. Soc.* **84** 2763-2769

Hoering T.C. & Parker P.L. (1961) The geochemistry of the stable isotopes of chlorine. *Geochim. Cosmochim. Acta* **23** 186-199

Kaufmann R.S. (1984) *Chlorine in groundwater: Stable isotope distribution.* Ph.D. Thesis, University of Arizona. 137 pp.

Kohnen M.E.L. (1988) *Stabiele chloorisotopen onderzoek.* Internal report. University of Utrecht. 17 pp.

Langvad T. (1954) Separation of chlorine isotopes by ion-exchange chromatography. *Acta Chem. Scand.* **8** 526-527

McIlvaine T.C. (1921) A buffer solution for colorimetric comparison. *J. Biol. Chem.* **49** 183-186

Owen H.R. & Schaeffer O.A. (1955) The isotope abundances of chlorine from various sources. *J. Amer. Chem. Soc.* **77** 898-899

Taylor J.W. & Grimsrud E.P. (1969) Chlorine isotopic ratios by negative ion mass spectrometry. *Anal. Chem.* **41** 805-810

Vogel A.I. (1951) *A textbook of quantitative inorganic analysis, theory and practice.* Longmans, Green and Co. London. 918 pp.

CHAPTER 3

Accuracy of the Mass Spectrometer for ^{37}Cl/^{35}Cl Ratios and the Method of CH$_3$Cl Preparation

H.G.M. Eggenkamp[1]

ABSTRACT-- The accuracy of δ^{37}Cl measurements is determined for the analytical procedures used in this study. Over a very long period the mass spectrometer is very stable. The δ^{37}Cl difference between two reference gasses remained virtually the same for a period over three years. A reference sample is also measured over the same period and this also gave virtually the same values over this period. Although samples can be measured with standard deviations less than 0.1‰, tests show that for a reliable determination the samples should be measured at least two times. A sample has to contain an amount of chloride equivalent to 100 μl seawater (1.9 mg Cl⁻), otherwise δ^{37}Cl values become of no value. Large samples seem to have no effect on the δ^{37}Cl although care must be taken to add enough chemicals to make reactions go to completion.

INTRODUCTION

Since natural δ^{37}Cl variations are small, it is important to know the accuracy of the methods. Reference gases were measured frequently during the whole period of this study in order to test the long term stability of the mass spectrometer. Sample preparation was tested by frequent analysis of a seawater reference sample.

FRACTIONATION IN THE MASS SPECTROMETER

The mass spectrometer has the possibility to remeasure a portion of gas af the standard deviation is larger than a previously set maximum value. A too large standard deviation can result, for example, when the change-over valve did not engage at the proper moment. Generally the maximum number of attempts to reach a standard deviation is two. If this standard deviation is set to zero and the maximum number of measurements is set to a high value (10, 20 or more) a portion of gas will be (re)measured that many times. This results in a repeated measurement of the same portion of gas. Each time a small portion of it (about 6 to 7% of the present quantity) is measured.

The following tests illustrate the fractionation of a distinct portion of gas within the mass spectrometer. On December 14, 1988 δ^{37}Cl decreased by 0.10‰ in ten repeat measurements. On December 16, 1988 δ^{37}Cl varied 0.11‰ within twenty measurements. No trend was observed. On October 26, 1991 two portions of CH$_3$Cl ware remeasured ten times, total differences were 0.09 and 0.11‰. On October 27th, one portion of gas was remeasured 33 times. After ten attempts the δ^{37}Cl value started to increase. The last measurement was 2.64‰ higher than the first (FIG. 1).

These results show that, although sometimes a trend may be present, the total

1Department of Geochemistry, Utrecht University, P.O.Box 80.021, 3508 TA Utrecht, The Netherlands

variations are small, if the number of repeat measurements is not too large (<10-20). Because the gas pressure in the mass spectrometer becomes very small fractionation will exist in each measured portion of gas. The residual fraction of the sample than also is fractionated and the effect would become more pronounced in the later measurements.

Fig. 1: Fractionation in the mass-spectrometer during repeated measurements of a distinct portion of CH₃Cl. ● = December 14, 1988; ♦ = December 16, 1988; ■ and ▲ = October 26, 1991; ♣ = October 27, 1991.

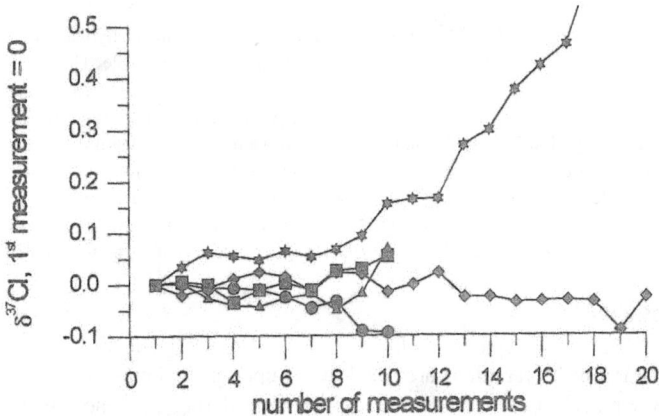

DECREASING AMOUNTS OF MEASURED GAS

For this test gas vessels were filled with pure CH_3Cl. After each measurement a fresh amount of gas was introduced into the mass spectrometer. On December 17, 1988 a gas vessel was measured 16 times. The first six measurements the transducer pressure decreased from 3.1 to 0.0 mbar. Till the eighth measurement enough gas was present to reach a major beam of $2.5*10^{-10}$A. After this measurement the major beam decreased by about 23% each measurement. Although the first measurements reached a major beam of $2.5*10^{-10}$A the measured $\delta^{37}Cl$ decreased by 0.26‰. When the major beam decreased $\delta^{37}Cl$ decreased quicker. The last measurable fraction (with a major beam of $2.78*10^{-11}$A) has a $\delta^{37}Cl$ 1.94‰ lower than the initial value. The test on October 26, 1991 showed similar results. The vessel was filled with 800 mbar of CH_3Cl so the transducer pressure was very high. It decreased from 30.9 to 0 in 12 measurements. $\delta^{37}Cl$ decreased by 0.19‰. After sixteen measurements the major beam was lower than $2.5*10^{-10}$A. Here $\delta^{37}Cl$ had decreased by 0.65‰ (see Fig. 2).

In all test runs $\delta^{37}Cl$ of the CH_3Cl decreased. Thus fractionation occurred on the manifold or in the mass spectrometer. Most probably it took place because the time that the sample moved from the manifold to the mass spectrometer was too short. Thus, the

24

Fig. 2: Decrease of measured $\delta^{37}Cl$ when fresh portions of CH_3Cl are introduced to the mass-spectrometer from the same gas vessel. ■ $= December\ 17,\ 1988;$ ● $= October\ 26,\ 1991.$

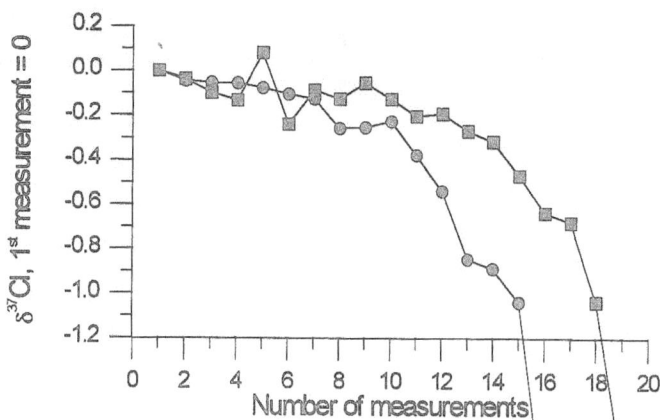

settings of the mass spectrometer are so that measurements can be done quickly, but a sample must not be measured too many times.

LONG TERM STABILITY OF THE MASS SPECTROMETER

Four 1-liter bottles made from glass were filled with CH_3Cl and analyzed frequently in order to monitor the long-term stability of the $^{37}Cl/^{35}Cl$ measurements. The gasses in reference bottles 1, 2, 3 and 4 were from lecture bottles obtained from Union Carbide®. Mass-spectrometric analysis involves comparison of the sample with a reference gas. Till May 24, 1989 reference 2 was used as the mass spectrometer reference gas, after that date reference 4 was used. This change was necessary because the gas pressure in reference bottle 2 had become too low and some air had leaked in. Reference 3 is the most frequently analysed gas (over the whole period of this study at least once before and after each batch of samples, total of 245 measurements). FIG. 3 shows the results on reference 3 plotted against the date (0 = April 30, 1988). The white dots refer to measurements relative to reference 2, the black ones to measurements relative to reference 4.

A striking point in FIG. 3 is that $\delta^{37}Cl$ relative to reference 2 decreases with time, whereas the measurements relative to reference 4 remained constant. This effect is probably due to the low pressure in reference bottle 2, causing isotope fractionation upon letting the gas into the mass spectrometer. The CH_3Cl pressure in reference bottles 1, 3 and 4 was much higher. Analyses were corrected for the changing isotopic composition of reference 2 based on a regression anlysis. Data measured relative to reference 4 were not corrected.

In **table 1** the most important general statistics of the four reference gasses can be found, for reference gas 1 relative to reference gas 2, for the other ones relative to reference gas 4. In all gasses the standard deviation is 0.06 or 0.07‰. This value is

Fig. 3: All measurements of reference gas 3. Open circles denote measurements relative to reference gas 2, dots denote measurements relative to reference gas 4.

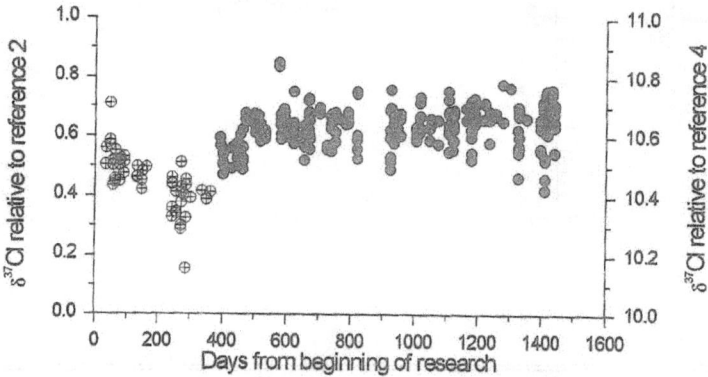

Table 1: *Statistics of the reference gasses. Values for reference 1 are relative to reference 2, all others relative to reference 4.*

	reference 1	reference 2	reference 3	reference 4
number	61	64	245	50
average	5.51	10.11	10.62	-0.01
median	5.52	10.11	10.63	0.00
standard deviation	0.06	0.06	0.07	0.07
minimum	5.32	9.98	10.40	-0.19
maximum	5.71	10.31	10.86	0.10
lower quartile	5.48	10.08	10.59	-0.06
upper quartile	5.53	10.13	10.67	0.04

assumed to be the typical error associated with the mass spectrometer. As can be seen from the total range and the interquartile range it is possible to measure relative large differences in one sample. This is shown in FIG. 4 where a histogram of all measured

Table 2: Variations of the reference gasses and Madeira 82 for each three months of this research. 88/2 to 89/2-1 are ‰ differences relative to reference gas 2, 89/2-2 to 92/2 are relative to reference gas 4. Per reference gas and per three months the number of measurements, the average and the standard deviation are given.

quarter	ref. 1		ref. 2		ref.3		ref. 4		seawater	
88/2	11	-4.58±0.05	13	0.13±0.04	12	0.53±0.07	-	-	6	-5.45±0.13
88/3	17	-4.61±0.05	21	0.08±0.02	14	0.48±0.03	-	-	4	-5.31±0.15
88/4	32	-4.75±0.05	12	0.00±0.06	8	0.38±0.10	-	-	24	-5.57±0.12
89/1	-	-	12	0.03±0.09	17	0.40±0.07	-	-	13	-5.58±0.13
89/2-1	-	-	-	-	3	0.40±0.01	10	-10.09±0.10	2	-5.55±0.04
89/2-2	-	-	4	10.09±0.02	9	10.53±0.04	1	-0.07	8	4.16±0.09
89/3	-	-	-	-	28	10.60±0.06	7	-0.01±0.05	29	4.29±0.09
89/4	-	-	-	-	19	10.63±0.03	2	-0.08±0.00	5	4.23±0.10
90/1	-	-	-	-	27	10.63±0.06	-	-	12	4.23±0.09
90/2	-	-	-	-	18	10.64±0.04	2	-0.07±0.01	7	4.18±0.07
90/3	-	-	-	-	6	10.63±0.10	-	-	1	4.09
90/4	-	-	-	-	25	10.61±0.09	4	-0.08±0.05	9	4.20±0.08
91/1	-	-	-	-	16	10.64±0.04	-	-	7	4.20±0.13
91/2	-	-	-	-	20	10.64±0.05	8	-0.01±0.06	19	4.15±0.07
91/3	-	-	-	-	23	10.68±0.04	12	0.04±0.04	14	4.22±0.04
91/4	-	-	-	-	16	10.66±0.07	8	-0.04±0.06	2	4.10±0.13
92/1	-	-	-	-	47	10.68±0.04	10	-0.02±0.07	36	4.17±0.06
92/2	-	-	-	-	11	10.68±0.03	4	-0.02±0.04	14	4.17±0.08

values from reference gas 3 can be found. It is therefore recommended to measure a sample always at least twice.

The variation over the time is determined by calculating the average values per quarter of a year (**table 2**). It is clear that in the first year of the study (when measurements were done relative to reference gas 2) measured values decreased. Later measured values (relative to reference 4) were constant.

REPEATED ANALYSES OF A SEAWATER REFERENCE SAMPLE

Stable isotope ratios are reported relative to a standard. For chlorine isotopes the standard is average seawater chloride (SMOC, Standard Mean Ocean Chloride). The oceans represent a very large and well mixed chloride reservoir. Therefore, the $\delta^{37}Cl$ of it is constant (KAUFMANN 1984). In our laboratory we used a sample of seawater from the Atlantic Ocean near Madeira, collected in 1982. This sample (Madeira 82) was analyzed frequently during the whole study. The resulting variations are a combination of mass spectrometer stability and sample preparation effects. In FIG. 5 all measured $\delta^{37}Cl$ values of Madeira 82 are shown the same way as they are for reference gas 3.

During the period that the measurements were done, the experimental method improved. This can be seen from **table 3** where the standard deviations of subsequent periods are 0.13, 0.14, 0.09 and 0.06‰. A major improvement of the accuracy was obtained when it was found that the temperature of the reaction between AgCl and CH_3I must be kept below 80 °C in order to avoid partial decomposition of CH_3I. The standard deviation of the seawater samples measured during the last period of this study is only slightly higher than that obtained on the reference gasses, indicating that the contribution by sample preparation is preparation small. In the earlier measurements the standard deviation is about 0.06‰ higher.

Fig. 4: Frequency histogram of all measurements of reference gas 3 relative to reference gas 4.

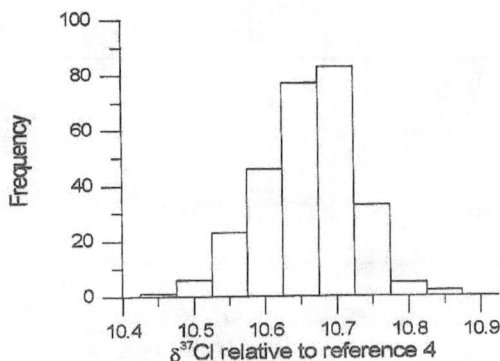

Comparing Figs. 3 and 5 a striking distinction can be made. In both figures the difference between the two y-axes is 10‰. The difference between the values measured relative to the reference gasses 2 and 4 is not equal. For Madeira 82 this difference is about 9.8‰ and for reference gas 3 it is about 10.2‰. We do not know the reason for this.

Table 3: *Statistics of the seawater standard (Madeira 82) for different periods.*

	relative to reference 2 (1988-1989)	relative to reference 4 (1989)	relative to reference 4 (1990-1991)	relative to reference 4 (1992)
number	43	53	40	49
average	-5.54	4.29	4.19	4.17
median	5.54	4.29	4.18	4.17
standard deviation	0.13	0.14	0.09	0.06
minimum	-5.80	4.00	4.03	4.05
maximum	-5.28	4.60	4.35	4.35
lower quartile	-5.63	4.21	4.13	4.14
upper quartile	-5.43	4.36	4.24	4.20

As an example of a factor which influences the $\delta^{37}Cl$ measurement the ratio between the ion gauge readings for the sample and the reference of the mass spectrometer is calculated. This ratio must be 1, but if a sample is contaminated, for example with water or reaction products of CH_3I the reference reading increases, and consequently this ratio.

Fig. 5: All measurements of Madeira 82. Circles are relative to reference gas 2, dots relative to reference 4.

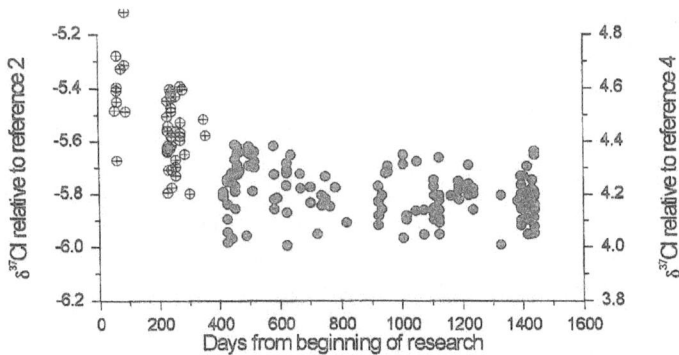

Fig. 6: $\delta^{37}Cl$ of Madeira 82 relative ref. 4 as a function of the ion gauge ratio.

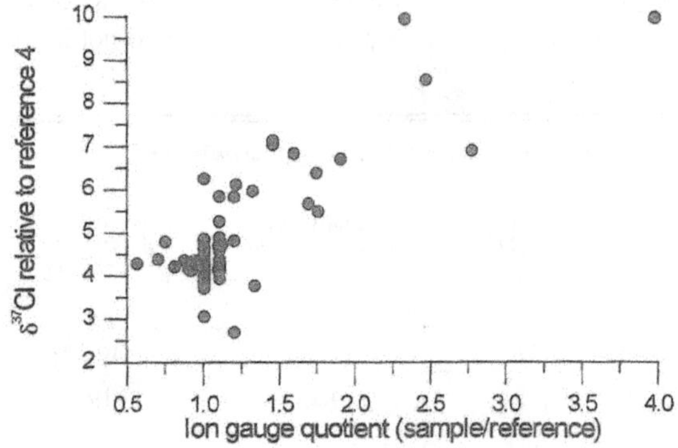

Fig. 7: AgCl yield in % of the expected yield.

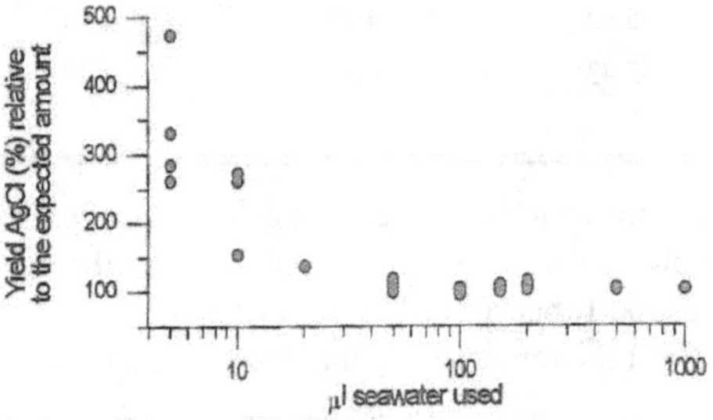

As can be seen in FIG. 6 the $\delta^{37}Cl$ changes dramatically and it is clear that the measured values have no meaning. These values are omitted from the data file.

Other samples that were discarded include samples which contained water (because of analytical errors) and samples from which some CH_3Cl is escaped. These samples are characterized by anomalous $\delta^{37}Cl$ values and also omitted from the file.

Average values for $\delta^{37}Cl$ of the seawater were determined every three months (see **table 2**). Because small differences are found between various periods, a moving average of the Madeira 82 results is taken as the reference value for the $\delta^{37}Cl$ measurements. All the data reported in the following chapters were calculated relative to this moving average, which is assumed to be equal to SMOC.

EFFECT OF SAMPLE SIZE

A series of experiments were made to test the effect of sample size on the obtained

δ^{37}Cl. Different amounts of Madeira seawater, ranging from 5 to 1000 µl, were measured. For small samples a relatively high influence of CH_3I decomposition can be expected, whereas large samples produce a thick layer of AgCl which may not react completely with CH_3I.

Table 4: *Influence of the amount of added seawater on the masured δ^{37}Cl.*

µl seawater	average	standard deviation	number of measurements
5	3.70	0.43	4
10	4.21	0.58	3
20	3.93	0.08	2
50	4.00	0.19	5
100	4.18	0.08	4
150	4.24	0.11	7
200	4.24	0.06	9
500	4.20	0.21	2
1000	4.19	0.03	2

It was found that for samples below 50µl , the AgCl yield obtained by weighing the glass fibre filter is much higher than expected (FIG. 7). It is suspected that contamination with dust during filtration is the source of this discrepancy.

Sample sizes below 100 µl tend to give δ^{37}Cl values that are too low and less

Fig. 8: *δ^{37}Cl of Madeira 82 as a function of the amount used.*

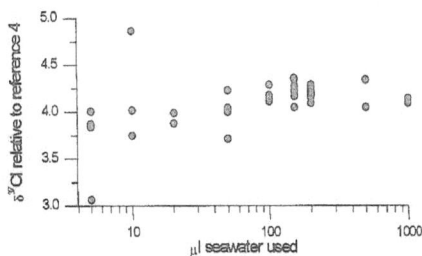

reproducible (**table 4** and FIG. 8). Below 20µl, the date become useless (**table 4**). From these results it is recommended to use at least the equivalent of 100 µl seawater, which is about 1.9 mg or $5*10^{-5}$ mole of chloride.

Large sample sizes up to 1000µl do not seem to have an effect on the AgCl yield nor on δ^{37}Cl, as long as more silver nitrate solution is added and more iodomethane.

Otherwise, isotope fractionation may occur because of incomplete precipitation or incomplete reaction.

REFERENCES

KAUFMANN R.S. (1984) *Chlorine in ground water: stable isotope distribution.* Ph.D. Thesis, University of Arizona. 136 pp.

CHAPTER 4

Theoretical Fractionation of Chlorine Isotopes During Molecular Diffusion

H.G.M. Eggenkamp[1]

ABSTRACT-- The theoretical fractionation of chlorine isotopes is calculated for four simple diffusion models, in which no sorption is assumed. Theoretically large fractionations can exist in pure chlorine diffusion systems. Because of the large diffusion coefficient of the conservative chloride ion the isotope signal flattens out relatively quickly in systems with a scale that is typical for sediment cores in geological studies. For different diffusion types, the shape of the $\delta^{37}Cl$ curve is sufficiently different so that it can used to evaluate the type of diffusion that has occurred.

INTRODUCTION

Isotopes fractionate during diffusion. SENFTLE & BRACKEN (1955) show that it is likely that isotope fractionation must occur in all diffusion systems, with the diffusion coefficients being higher in liquids and gases than in rocks. The geochemical behaviour of chlorine is very conservative, for it interacts little with its surroundings. Also, under most natural conditions the chloride ion is the only existing oxidation state, so that redox reactions do not affect the isotopic composition. Therefore, diffusion is the main process to produce chlorine isotope fractionation.

DESAULNIERS *et al.* (1986) showed that diffusion caused $\delta^{37}Cl$ variations in ground water systems in Ontario, where a correlation was found between the chloride concentration and $\delta^{37}Cl$. The chlorine isotope variations occur because the heavier isotope ^{37}Cl has a slightly smaller diffusion coefficient than the lighter isotope ^{35}Cl. MADORSKY & STRAUSS (1947) and KONSTANTINOV & BAKULIN (1962) determined that the diffusion coefficient ratio of the two isotopes lies between 1.0012 and 1.0022. This range is rather large because the analytical methods at that time were not as good as they are now. In chapter 5 of this thesis it will be shown that this ratio is 1.0023 in pore waters from Kau Bay, Halmahera, Indonesia.

The theory of diffusion of radioisotopes was treated by DUURSMA & HOEDE (1967) who presented mathematical solutions for several models of diffusion. For the calculation of chlorine isotope fractionation, we used four different models describing diffusion without sorption:

1) Diffusion from a source having a constant concentration.

2) Diffusion from a source having a constant concentration, combined with sedimentation or advective flow of pore water.

3) Diffusion from a momentary release, and

4) Diffusion from a source with a constant inflow.

---

[1]Department of Geochemistry, Utrecht University, P.O.Box 80.021, 3508 TA Utrecht, The Netherlands

DIFFUSION MODELS

Molecular diffusion is the process in which matter is transported from one part of a system to another as a result of arbitrary molecular movements (CRANK 1956). It was first described by Adolf Fick in 1855. This work is now referred to as Fick's First and Second Law, and was published even before quantitative experimental measurements had been done.

According to Fick's First Law the amount of matter (dm) moving through a plane perpendicular to the direction of diffusion during a time (dt) is proportional to the concentration difference on both sides of the plane and the area of the plane.

$$\frac{\partial m}{\partial t} = -DA\frac{\partial c}{\partial x} \tag{1}$$

D is the diffusion coefficient, A is the area of the plane, c is the concentration and x is the migration distance.

The amount of matter moving through the plane per unit of time is called the flux (j):

$$j = \frac{\partial m}{A\partial t} \tag{2}$$

From which it follows that:

$$j = -D\frac{\partial c}{\partial x} \tag{3}$$

Which is known as the mathematical formulation of Fick's First Law. The diffusion coefficient D gives the amount of matter that moves through a unit of area in a unit of time in a unit of concentration gradient. Its dimension is [length]^2/[time].

It is not possible to measure the value of dm/dt in equation (1) directly. This is adressed by Fick's Second Law, which can be derived from the First Law and the Law of preservation of matter. The second law defines that if more matter is supplied than removed the concentration increases, and vice versa.

$$\frac{\partial c}{\partial t} = D\frac{\partial^2 c}{\partial x^2} \tag{4}$$

$$\frac{\partial c}{\partial t} = D\left(\frac{\partial^2 c}{\partial r^2} + \frac{1}{r}\frac{\partial c}{\partial r}\right) \tag{5}$$

$$\frac{\partial c}{\partial t} = D\left(\frac{\partial^2 c}{\partial r^2} + \frac{2}{r}\frac{\partial c}{\partial r}\right) \tag{6}$$

This presents Fick's second law for linear (one-dimensional, equ. 4), cylindrical (two-dimensional, equ. 5) and spherical (three-dimensional, equ. 6) diffusion respectively. In this equation x or r represents the distance to the diffusion plane or centre. In the latter two cases it is supposed that the concentration is radially symmetric and that with respect to the origin, the diffusion is isotropic.

-Diffusion from a source with a constant concentration

An infinite amount of matter with high chloride content diffuses into an infinite amount of matter with low chloride content. The chloride concentration on the boundary between these two parts will be constant during diffusion. Only the linear diffusion version produces a realistic model; in the case of cylindrical or spherical diffusion the source at $r=0$ is either a line or a point and the constant concentration at the source cannot be maintained at any time $t>0$ (Duursma & Hoede 1967).

The solution for the concentration as a function of x and t is (see e.g. Carslaw & Jaeger 1959):

$$c(x,t)=c_0 erfc \frac{x}{2\sqrt{Dt}}$$ (7)

in which "erfc" is the complementary error function, defined as:

$$erfc\, z = \frac{2}{\sqrt{\pi}} \int_z^\infty e^{-y^2} dy$$ (8)

In our calculations the complementary error function is approximated with:

$$erfc\, z = (a_1 y + a_2 y^2 + a_3 y^3 + a_4 y^4 + a_5 y^5) e^{-z^2} + \epsilon(z)$$

$$y = \frac{1}{1+pz}$$

$$|\epsilon(z)| \leq 1.5 \times 10^{-7}$$ (9)

$$p = 0.3275911,\ a_1 = 0.254829592$$
$$a_2 = -0.284496736,\ a_3 = 1.421413741$$
$$a_4 = -1.453152027,\ a_5 = 1.061405429$$

(from Hastings Jr. 1955, see Abramowitz & Stegun 1968). Parameters p and a_1 to a_5 are determined in a numerical and empirical nature (Hastings Jr. 1955).

-Diffusion from a source with a constant concentration, combined with advective pore water flow

The diffusion model described above is extended with an extra component for pore water advection (Middelburg & de Lange 1989, Middelburg 1990). The diffusion equation then becomes:

$$\frac{\partial c}{\partial t} = D\left(\frac{\partial^2 c}{\partial x^2}\right) - V\left(\frac{\partial c}{\partial x}\right)$$ (10)

where V is the pore water advection rate.

Following Middelburg & de Lange (1989), we assume that at $t=0$ the chloride concentration in the pore water is constant with depth, whereas at $t>0$ the chloride concentration at the water-sediment interface is constant and as $t>0$ and depth approaches infinity $\partial c/\partial x=0$. The solution of this equation is:

$$c = c_i + (c_0 - c_i)B(x,t)$$
$$where$$
$$B(x,t) = \frac{1}{2} erfc\left[\frac{(x-Vt)}{2\sqrt{Dt}}\right] + \frac{1}{2}\exp\left(\frac{Vx}{D}\right) erfc\left[\frac{(x+Vt)}{2\sqrt{Dt}}\right]$$ (11)

-Diffusion after a momentary release of chloride

In this model, at time $t = 0$ a distinct amount of chloride ions is liberated. From the point or line where x or $r = 0$ the chloride will diffuse into the direction of the lower concentration. The concentration at x or $r = 0$ will decrease as t increases. For this model, the solutions for the three possible cases, linear (equ. 12), cylindrical (equ. 13) and spherical (equ. 14) diffusion are (JOST 1957):

$$c(x,t)=\frac{s}{(4\pi Dt)^{1/2}}\exp\left(-\frac{x^2}{4Dt}\right) \tag{12}$$

$$c(r,t)=\frac{s}{4\pi Dt}\exp\left(-\frac{r^2}{4Dt}\right) \tag{13}$$

$$c(r,t)=\frac{s}{(4\pi Dt)^{3/2}}\exp\left(-\frac{r^2}{4Dt}\right) \tag{14}$$

Where s is the amount of the momentary released chloride.

-Diffusion from a source with a constant inflow

In this model, a constant inflow of chloride with an infinitely small volume is liberated from a source at a constant rate. It can be considered as an extension of the former model. The diffusion solution of this model can be found by integrating equations 12, 13 and 14 over time, replacing s by qdt, (see CARSLAW & JAEGER 1959). In the solutions describing the diffusion from a constant inflow q defines the amount of chloride added per unit of time.

For linear diffusion the solution is given by (CARSLAW & JAEGER 1959):

$$c(x,t)=\frac{qx}{2D\sqrt{\pi}}\left[\frac{2\sqrt{Dt}}{x}\exp\left(-\frac{x^2}{4Dt}\right)-\sqrt{\pi}\,erfc\frac{x}{2\sqrt{Dt}}\right] \tag{15}$$

The solution for the two-dimensional (cylindrical) diffusion is (CARSLAW & JAEGER 1959):

$$c(r,t)=\frac{-q}{4\pi D}Ei\left(\frac{-r^2}{4Dt}\right) \tag{16}$$

in which Ei stands for the "exponential integral" which is defined as:

$$Ei(-z)=\int_z^\infty \frac{e^{-y}}{y}dy \quad (|argz|<\pi) \tag{17}$$

For $0 \leq z \leq 1$ the "exponential integral" is approximated with:

$$Ei(z)+\ln z = a_0+a_1z+a_2z^2+a_3z^3+a_4z^4+a_5z^5+\epsilon(z)$$
$$|\epsilon(z)<2\text{x}10^{-7}|$$
$$a_0=-0.57721566,\ a_1=0.99999193 \tag{18}$$
$$a_2=-0.24991055,\ a_3=0.05519968$$
$$a_4=-0.00976004,\ a_5=0.00107857$$

(from ALLEN 1954) and for $1 \leq z \leq \infty$ the "exponential integral" is approximated with:

$$ze^z Ei(z)=\frac{z^4+a_1 z^3+a^2 z^2+a_3 z+a_4}{z^4+b_1 z^3+b_2 z^2+b_3 z+b_4}+\epsilon(z)$$

$$|\epsilon(z)|<2\text{x}10^{-8}$$

$$a_1=8.5733287401,\ a_2=18.0590169730 \tag{19}$$

$$a_3=8.6347608925,\ a_4=0.2677737343$$

$$b_1=9.5733223454,\ b_2=25.6329561486$$

$$b_3=21.0996530827,\ b_4=3.9584969228$$

(from HASTINGS JR. 1955). Parameters a_1 to a_5 and b_1 to b_4 are determined in a numerical and empirical nature (HASTINGS JR. 1955).

The solution for spherical diffusion is (CARSLAW & JAEGER 1959):

$$c(r,t)=\frac{q}{4\pi rD}erfc\frac{r}{2\sqrt{Dt}} \tag{20}$$

-Calculation of $\delta^{37}Cl$ from concentration profiles

The $\delta^{37}Cl$ model values are calculated from the calculated concentrations of the isotopes ^{35}Cl and ^{37}Cl using the standard definition of the delta value. So the $\delta^{37}Cl$ is calculated as:

$$\delta^{37}Cl=\frac{\left(\frac{c_{37Cl}}{c_{35Cl}}\right)_{calculated}-\left(\frac{c_{37Cl}}{c_{35Cl}}\right)_{standard}}{\left(\frac{c_{37Cl}}{c_{35Cl}}\right)_{standard}}*1000 \tag{21}$$

$\delta^{37}Cl$ is expressed as the per mil (‰) deviation relative to the standard, which is by definition 0‰. The standard for chlorine isotopes is average seawater (Standard Mean Ocean Chloride, SMOC). The calculations were done using $\delta^{37}Cl=0$ as the overall starting composition of the system.

Thus the mathematical expressions developed by DUURSMA & HOEDE (1967), where the differences are determined using the different diffusion coefficients for both isotopes. Calculations are done in GWBASIC. For diffusion from a constant source, combined with sedimentation or advective flow of ground water, calculations by MIDDELBURG & DE LANGE (1989) are used (also in GWBASIC).

RESULTS AND DISCUSSION

-Diffusion from a source with a constant concentration

This first system consists of a layer of sediments of infinite thickness and saturated with seawater, lying on a layer of sediment, also with infinite thickness, saturated with either fresh or brackish water. In all calculations, the chloride concentration of seawater is taken as 540 mM (19145 ppm), the diffusion coefficient of chloride is $15*10^{-10}$ m^{2}s^{-1} and $\delta^{37}Cl$ of the system is 0‰ at time $t=0$. The calculations were done for varying diffusion times, varying concentrations in the fresh/brackish sediment layer and varying diffusion coefficient ratios. The conditions for which calculations have been carried out are given in **table 1**.

The effect of varying diffusion times is shown in FIG. 1. Diffusion profiles are calculated for 100, 250, 500, 1000 and 2500 years of diffusion. The chloride concentration at position $x = 0$ (boundary between the two layers) remains constant, but the concentration gradients become smaller with increasing time. As time approaches infinity the concentration will become the same in both layers, and equal to the average values at $t = 0$. $\delta^{37}Cl$ at position $x = 0$ remains zero. Because ^{35}Cl has a slightly higher diffusion coefficient than ^{37}Cl, it will migrate quicker into the fresh water layer. Thereby $\delta^{37}Cl$ of the seawater layer will increase, whereas $\delta^{37}Cl$ of the freshwater will decrease. At longer distances, where no chloride is added or removed, no isotope fractionation occurs

Table 1: *Diffusion conditions diffusion from a constant source.*

FIG.	time (years)	Initial freshwater concnetration (mM)	diffusion coefficient ratio
1	variable	5	1.00245
2	250	variable	1.00245
3	250	5	variable

and $\delta^{37}Cl$ will remain 0‰. So, the $\delta^{37}Cl$ profiles show a maximum in the sea water layer, and a minimum in the fresh water layer. Because the chloride concentration in the fresh water layer is low, the isotope effect is more apparent in the fresh water layer than in the sea water layer. For the same reason, the maximum in the seawater layer is closer to $x = 0$ than the minimum in the freshwater layer. With increasing time the extremes move away from the position $x = 0$. The maximum and minimum $\delta^{37}Cl$ that are reached, however, do

Fig. 1: *Diffusion from a constant source with variable diffusion times*

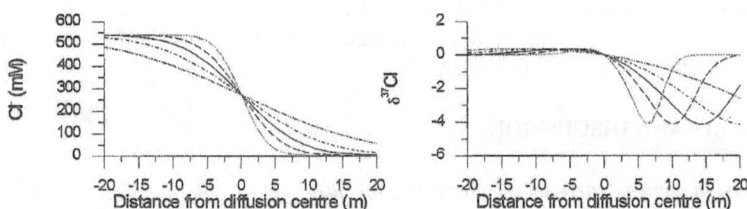

not change with time. Under the conditions chosen in FIG. 1, the maximum $\delta^{37}Cl$ is always +0.36‰ and the minimum is always -4.14‰

The chloride concentration of the fresh water layer has a large effect on $\delta^{37}Cl$, especially when this concentration is low. FIG. 2 shows diffusion profiles for initial freshwater concentrations of 0, 5, 25, 100 and 500 mM (0, 177, 886, 3545 and 17727 ppm). The other conditions are shown in table 1. The minimum value of $\delta^{37}Cl$ in the freshwater depends strongly on the initial chloride concentration. In the (theoretical) situation that the initial chloride content was zero, $\delta^{37}Cl$ will go to minus infinity because no initial chloride is present to buffer $\delta^{37}Cl$ at 0‰. For higher initial chloride contents, the minimum is found closer to $x = 0$ and $\delta^{37}Cl$ is closer to 0‰. The initial chloride concentration in the fresh water layer has only little effect on the maximum $\delta^{37}Cl$ in the seawater layer.

Changing the diffusion coefficient ratio (FIG. 3) affects only the values of the $\delta^{37}Cl$ maximum and the $\delta^{37}Cl$ minimum, not the chloride concentration profile. Shown are the profiles for $\alpha_{35/37} = 1.00295, 1.00245, 1.00195, 1.00145$ and 1.00095. In natural systems where all the other parameters are known, the chloride isotope profile can therefore be

Fig 2: Diffusion from a constant source with variable initial fresh water concentrations.

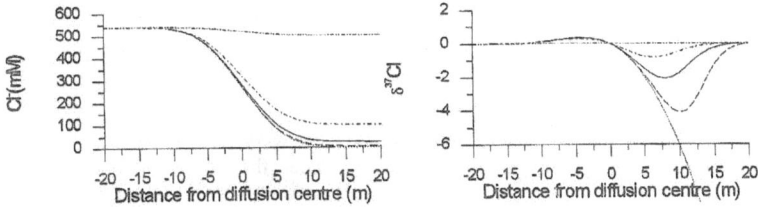

Fig. 3: Diffusion from a constant source with variable diffusion coefficient ratios.

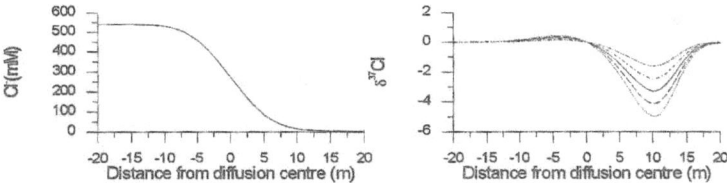

used to determine the chlorine isotope diffusion coefficient ratio, as will be shown in chapter 5.

-Diffusion from a source with a constant chloride concentration, combined with sedimentation or pore water advective flow

This model describes the situation where an infinite water body with constant chloride concentration (such as the ocean) lies on top of freshwater sediments. Sedimentation is modeled as a downward advective flow of pore water, equal to the sediment accumulation rate. If there is an additional upwards pore water flow through the sediments (e.g. as described by MIDDELBURG & DE LANGE (1989)), this appears as a negative contribution to the sediment accumulation rate. FIG. 4 shows the diffusion profiles after 100, 250, 500, 1000 and 2500 years, in the case that the sediment accumulation rate (or pore water advective rate) is 0.1 cm.a^{-1}. The other conditions are as in FIG. 1 (see **table 1**). As we have seen before (model <u>without</u> sedimentation), the concentration gradients become smaller with increasing diffusion times. And as we have also seen, the minimum δ^{37}Cl value does not change with time.

FIG. 5 shows the diffusion profiles calculated for a diffusion time of 250 years and sedimentation/advection rates of +5, +2.5, 0, -2.5 and -5 cm.yr^{-1}. High sedimentation rates result in a concentration profile that has a relatively flat upper part. This is because the high sedimentation rates leave relatively little time for diffusion. With lower values this "platform" disappears, and a steep slope is found at the sediment-water interface. Two striking effects can be found in δ^{37}Cl profiles. The first is a relation between the minimum δ^{37}Cl and the advection rate, and the second are the slightly positive δ^{37}Cl values in systems with a positive effective advection rate. The change in minimum δ^{37}Cl can be explained as follows: as the advection rate is negative, a net upwards flow of fresh water appears to the sediment-water surface, where concentration is always constant. Chloride diffusion into the sediment is forced back. ^{35}Cl diffuses easier into the sediment, causing a negative δ^{37}Cl. The back force pushes the chloride back, where again ^{35}Cl moves back easier, resulting in an extra low δ^{37}Cl.

The positive δ^{37}Cl values that occur at positive advection rates (= sedimentation) occur because this situation is somewhere in-between the situation where well-mixed

Fig. 4: Diffusion from a constant source, combined with a pore water advection rate of 0.1 cm/yr with variable diffusion times.

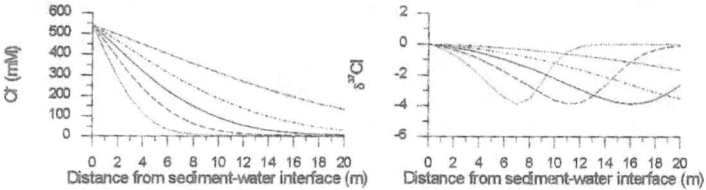

Fig. 5: Diffusion from a constant source, combined with variable pore water advection rates.

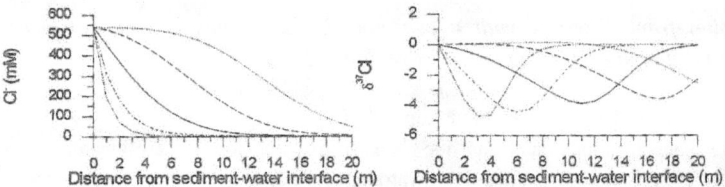

seawater lies on top of freshwater sediments and the situation where an infinite layer of marine sediments lies on top of fresh water sediments.

-Diffusion after a momentary release of chloride

In this model, a limited amount of chloride diffuses in one (linear), two (cylindrical) or three (spherical) dimensions. Diffusion profiles were calculated for a fixed amount of chloride (10 mole) and diffusion times of 100, 250, 500, 1000 and 2500 years. The results are shown in FIGS. 6 (linear diffusion), 7 (cylindrical diffusion) and 8 (spherical diffusion). Calculations were also done for a fixed diffusion time (250 years) and various amounts of chloride (1, 3, 10, 30 and 100 mole). The results are shown in FIGS. 9, 10 and 11. In all cases, an initial concentration of 5 mmole is assumed in the sediment surrounding the diffusion centre. The diffusion coefficient in all calculations is 15.10^{-10} $m^2.s^{-1}$ and the diffusion coefficient ratio is 1.00245.

With increasing time, chloride diffuses away from the source, and therefore the concentration decreases (FIGS. 6, 7 and 8). This effect is more pronounced in a two dimensional than in a one dimensional system, and more pronounced in a three dimensional than in a two dimensional system. The reason for this is that the same amount of chloride has to be spread over more directions and consequently will be more diluted. With increasing time concentration at the diffusion centre decreases because the remaining amount of chloride in the centre, which was fixed will be lowered by loss through diffusion. The $\delta^{37}Cl$ profiles for one-, two-, and three-dimensional diffusion have similar shapes. However, because the concentration gradients are flatter in the higher dimensional systems the $\delta^{37}Cl$ profiles are also flatter. $\delta^{37}Cl$ at the source is higher in the more dimensional systems, because a larger proportion of the chloride diffused away so that ^{35}Cl was more effectively separated from ^{37}Cl.

The effect of different amounts of chloride (1, 3, 10, 30 and 100 mole) is shown in FIGS. 9, 10 and 11. The concentration profiles are simple: higher amounts of chloride produce higher chloride concentrations. The $\delta^{37}Cl$ profiles show that the isotope effects become larger when larger amounts of chloride are released; this is a result of the larger concentration gradients between the source and the surrounding sediments.

It appears from FIGS. 6 to 11 that the chloride concentration and $\delta^{37}Cl$ at the source vary with time and with the initial amount of chloride in the source. Variations at the source as a function of time are shown in FIG. 12. Chloride concentration and $\delta^{37}Cl$ in the source is shown as a function of time for an initial chloride source of 10 mole and a chloride concentration in the sediment of 5 mM. For all three diffusion types chloride concentration in the centre decreases with increasing time. This decrease is linear in a log-log plot. As the concentration approaches the initial concentration in the sediment it will become equal to it. This means that the effect of the initial source will end as the concentration is low. This is also found in the $\delta^{37}Cl$ plot. $\delta^{37}Cl$ is constant and high when diffusion times are short, and $\delta^{37}Cl$ for the higher dimensional diffusion is higher than for the lower dimensional systems. As the chloride concentration approaches the value of the chloride in the original sediment $\delta^{37}Cl$ will also approach $\delta^{37}Cl$ of the original sediment,

Fig. 6: Linear diffusion from a momentary release of chloride with variable diffusion times.

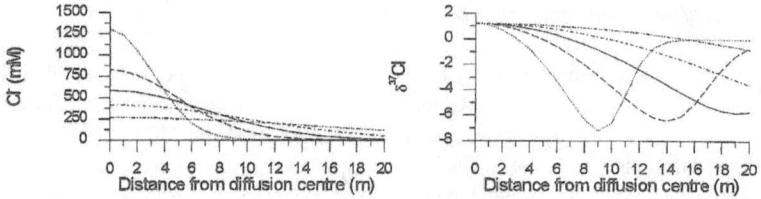

Fig. 7: Cylindrical diffusion from a momentary release of chloride with variable diffusion times.

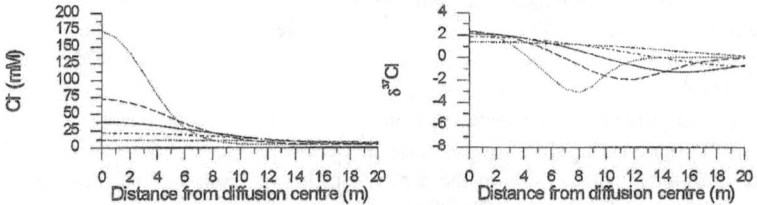

Fig 8: Spherical diffusion from a momentary release of chloride with variable diffusion times.

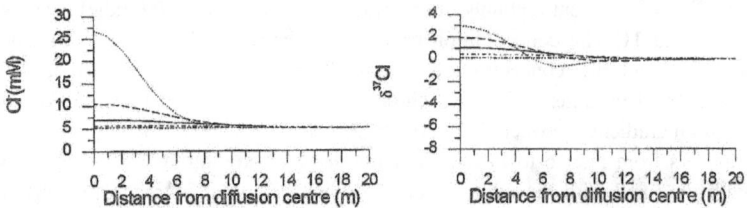

i.e. 0‰.

In FIG. 13 the variation in chloride concentration and $\delta^{37}Cl$ at the source are plotted as a function of the initial amount of chloride for a diffusion time of 250 years. When the amounts of chloride are very small, no effect is seen at the source. With larger amounts of chloride, the concentration at the source varies linearly (on a log-log plot). A similar effect is found for $\delta^{37}Cl$; when the amount of chloride is very small $\delta^{37}Cl$ remains zero. When the amount of chloride is larger, $\delta^{37}Cl$ increases and becomes independent of the amount of chloride.

-Diffusion from a constant inflow

Fig. 9: Linear diffusion from a momentary release of chloride with variable momentary releases.

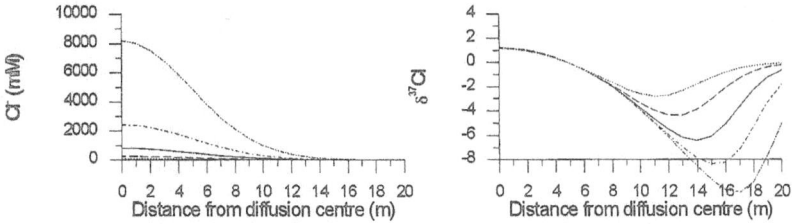

Fig. 10: Cylindrical diffusion from a momentary release of chloride with variable momentary releases.

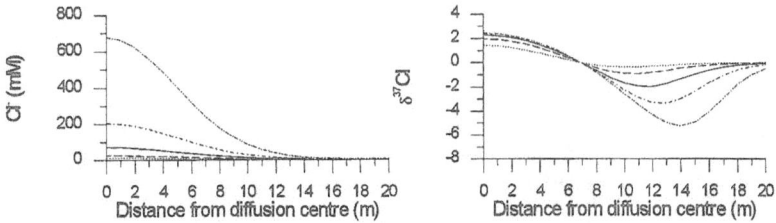

Fig. 11: Spherical diffusion from a momentary release of chloride with variable momentary releases.

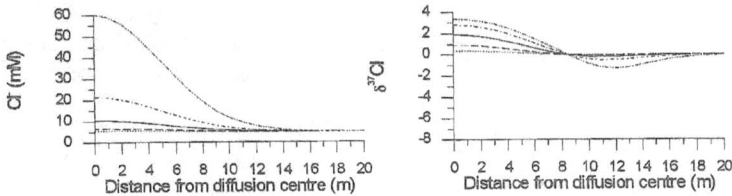

In this model, chloride is introduced at a constant rate and diffuses away from the source in one (linear), two (cylindrical) or three (spherical) dimensions. Again diffusion profiles are calculated for various diffusion times of 100, 250, 500, 1000 and 2500 years (Figs. 14, 15 and 16) and a fixed input of chloride (10 mmole/year) and for various input rates of 1, 3, 10, 30 and 100 mmole/year (Figs. 17, 18 and 19) for a fixed time of 250 years. In all these calculations an initial chloride concentration in the sediments of 5 mM is assumed, the diffusion coefficient is 15.10^{-10} $m^2.s^{-1}$ and the diffusion coefficient ratio

Fig. 12: Diffusion from a momentary release of chloride concentration and $\delta^{37}Cl$ at the source as a source of time.

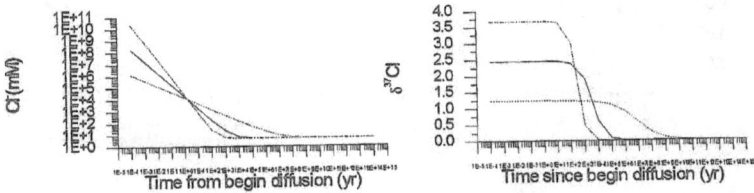

Fig. 13: Diffusion from a momentary release of chloride. Concentration and $\delta^{37}Cl$ at the source as a function of time.

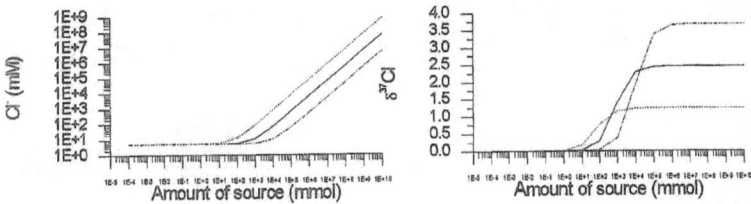

1.00245.

With a constant inflow of chloride, the concentration in any part of the system increases with time (Figs. 14, 15 and 16). For cylindrical and spherical diffusion, the concentration gradients close to the source are much steeper than for linear diffusion, because the chloride diffuses in more directions. As we have seen before, the minimum $\delta^{37}Cl$ is lower for linear diffusion than for cylindrical and spherical diffusion.

Diffusion profiles for input rates of 1, 3, 10, 30 and 100 mmole/year and a time of 250 years are shown in Figs. 17, 18 and 19. Both the chloride concentration profiles and the $\delta^{37}Cl$ profiles show similar effects as in the momentary release models. The explanations of these effects are the same.

No calculations were made for variations at the source. For x or $r = 0$, there is a problem of division by 0 in the equations for linear and cylindrical diffusion, and a factor $-\ln r$ in the equation for spherical diffusion.

CONCLUSIONS

Fig. 14: Linear diffusion from a constant inflow with variable diffusion times.

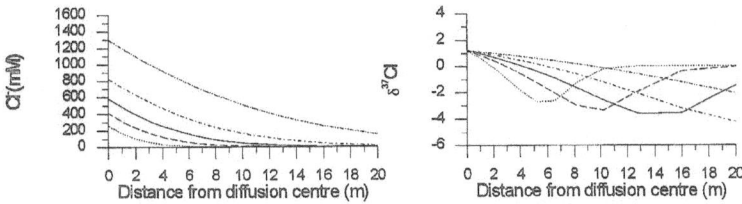

Fig. 15: Cylindrical diffusion from a constant inflow with variable diffusion times.

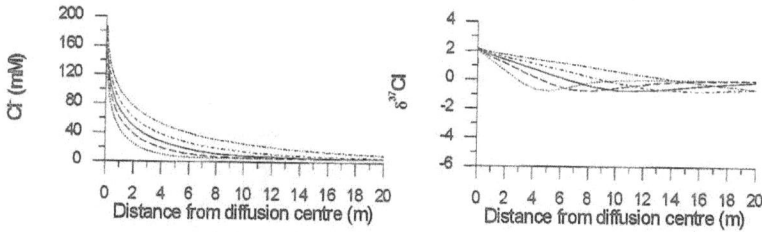

Fig. 16: Spherical diffusion from a constant inflow with variable diffusion times.

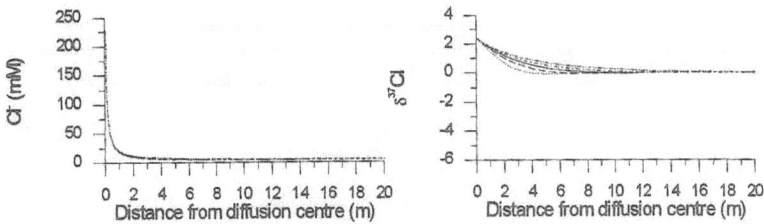

The model calculations show that diffusion is an effective mechanism to produce chlorine isotope variations. $\delta^{37}Cl$ values of -5‰ or even lower can be reached in natural diffusion systems under normal conditions. On the scale of typical sediment cores for geological studies, the isotope signal flattens out in a geologically short time (in the order of 10000 years); this is a result of the high diffusion coefficient of the chloride ion. Studying $\delta^{37}Cl$ variations in natural diffusion systems, the dimensions of the system must therefore be taken into account.

The shape of the $\delta^{37}Cl$ curves is different for different diffusion models. $\delta^{37}Cl$ profiles may give information on the type and boundary conditions of diffusion, especially when diffusion times are not too long, or the system is large.

Fig. 17: Linear diffusion from a constant inflow with variable amounts of constant inflow.

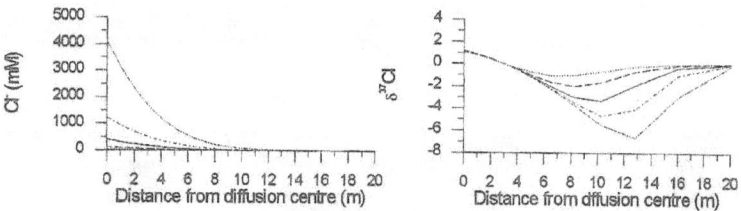

Fig. 18: Cylindrical diffusion from a constant inflow with variable amounts of constant inflow.

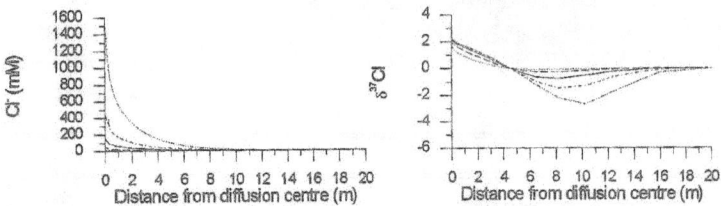

Fig. 19: Spherical diffusion from a constant inflow with variable amounts of constant inflow.

FIG. 19: Spherical diffusion from a constant inflow with variable amounts of a constant inflow.

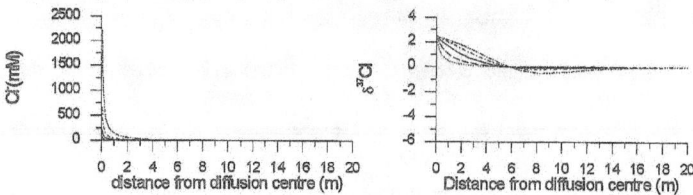

ACKNOWLEDGEMENTS

S.O. Scholten and R.Kreulen carefully read an earlier version of this paper and suggested many improvements. D.C. McCartney is thanked for linguistic advice.

REFERENCES

ABRAMOWITZ M. & STEGUN I.A. (1968) *Handbook of mathematical functions.* Dover Public., Inc., New York. 1043 pp.

ALLEN E.E. (1954) Analytical approximations. *Math. Tabl. Aid Comp.* **8** 240-241

CARSLAW H.S. & JAEGER J.C. (1959) *Conduction of heat in solids.* Oxford Univ. Press, London. Second Edt. 510 pp.

CRANK J.(1956) *The mathematics of diffusion.* Oxford University Press. 347 pp.

DESAULNIERS D.E., KAUFMANN R.S., CHERRY J.A. & BENTLEY H.W. (1986) ^{37}Cl-^{35}Cl variations in a diffusion controlled groundwater system. *Geochim. Cosmochim. Acta* **50** 1757-1764

DUURSMA E.K. & HOEDE C. (1967) Theoretical, experimental and field studies concnerning molecular diffusion of radioisotopes in sediments and suspended solid particles of the sea. Part A: Theories and mathematical calculations. *Neth. J. Sea Res.* **3** 423-457

HASTINGS JR. C. (1955) *Approximations for digital computers.* Princeton University Press. 201 pp.

JOST W. (1957) *Diffusion.* Verl. Dr. D. Steinkopf, Darmstadt. 177 pp.

KONSTANTINOV B.P. & BAKULIN E.A. (1965) Separation of chloride isotopes in aqueous solutions of lithium chloride, sodium chloride, and hydrochloric acid. *Russ. J. Phys. Chem.* **39** 315-318

MADORSKY S.L. & STRAUSS S. (1947) Concentration of isotopes of chlorine by the counter-current electromigration method. *J. Res. Nat. Bur. Stand.* **38** 185-189

MIDDELBURG J.B.M. (1990) Early diagenesis and authigenic mineral formation in anoxic sediments of Kau Bay, Indonesia. *Geol. Ultraj.* **71** 177 pp. Ph.D. Thesis, University of Utrecht.

MIDDELBURG J.J. & DE LANGE G.J.(1989) The isolation of Kau Bay during the last glaciation: direct evidence from interstitial water chlorinity. *Neth. J. Sea Res.* **24** 615-622

SENFTLE F.E. & BRACKEN J.T. (1955) Theoretical effect of diffusion on isotopic abundance ratios in rocks and associated fluids. *Geochim. Cosmochim. Acta* **7** 61-76

CHAPTER 5

Preferential Diffusion of ^{35}Cl Relative to ^{37}Cl in Sediments of Kau Bay, Halmahera, Indonesia

H.G.M. Eggenkamp[1], J.J. Middelburg[2] and R. Kreulen[1]

ABSTRACT-- Kau Bay, a 470 meter deep basin, is separated from the Pacific Ocean by a shallow, 40 meter deep sill. During the last glaciation this sill was above sea level, and the bay was turned into a fresh water lake. After glaciation, the sea level rose and the bay turned saline again. Since then, chloride from the newly formed saline sediments diffused into the freshwater sediments. As the sedimentation rate was constant throughout the Holocene it is relatively easy to model the chloride concentration in the sediment pore water. Because the diffusion coefficient of ^{35}Cl is slightly higher than the diffusion coefficient of ^{37}Cl, variations in δ^{37}Cl were expected in the pore water. Since the pore water history of Kau Bay is well known it was possible to use δ^{37}Cl variations to determine the diffusion coefficient ratio (D_{35}/D_{37}) for the two stable isotopes of chlorine. It was found that this ratio is 1.0023 in the pore water from sediments in this bay.

INTRODUCTION

Some decades ago, it was theoretically and experimentally shown that isotopes are fractionated by diffusion (e.g. MADORSKY & STRAUSS 1947). Isotope separation by diffusion results from the higher mobility of the light isotope relative to the heavy isotope. The isotope enrichment factor due to diffusion is defined as the ratio of the diffusion coefficients which reflect the ratio of ionic mobilities:

$$\alpha = \frac{D_L}{D_H} = \frac{U_L}{U_H} \tag{1}$$

where α is the isotope fractionation factor due to diffusion, D is the diffusion coefficient and U is the ionic mobility and the subscripts L and H refer to light and heavy isotopes respectively. For most other elements, isotope fractionation related to diffusion is masked by fractionation caused by chemical and biological processes. However, in sedimentary environments conservative elements, such as chlorine, are not significantly involved in biogeochemical processes and diffusion may, thus, be the major process that causes isotope fractionation.

In nature, the abundance ratio of the two stable isotopes of chlorine, ^{35}Cl and ^{37}Cl, is extremely constant. Deviations from the chlorine isotopic composition of seawater, expressed as δ^{37}Cl relative to Standard Mean Ocean Chloride (SMOC) are usually less than 1‰. As a consequence, no significant fractionation of chlorine isotopes was reported until the eighties (OWEN & SHAEFFER 1954; HOERING & PARKER 1961; MORTON & CATANZARO 1964). Recent improvements in sample preparation and mass spectrometry techniques allow us to determine these small natural variations in chlorine isotope ratios and to use

1Department of Geochemistry, Utrecht University, P.O.Box 80.021, 3508 TA Utrecht, The Netherlands
2Netherlands Institute of Ecology, Centre for Estuarine and Coastal Ecology, Vierstraat 28, 4401 EA Yerseke, The Netherlands

these variations to solve geological problems (e.g. KAUFMANN 1984, 1989; KAUFMANN *et al.* 1984, 1987, 1988; DESAULNIERS *et al.* 1986; EASTOE *et al.*, 1989; EASTOE & GUILBERT 1992; EGGENKAMP 1994).

The chlorine isotope fractionation factor, due to diffusion, can be obtained in three ways. Firstly, by analogy to gases the isotopic fractionation factor in aqueous solution can be estimated from the ratio of the square root of their masses (Graham's Law):

$$\alpha = \sqrt{\frac{M_H}{M_L}} \qquad (2)$$

where *m* is the molecular mass of the diffusing isotope. There are, however, two problems whth this approach: A) Collisions between gas molecules are elastic, whereas those between hydrated molecules and water molecules are inelastic. B) The masses of actual diffusing molecules in the solution must be known. This requires knowledge of the degree of chloride hydration, and the residence time of a water molecule next to a chloride ion relative to a water molecule next to another water molecule. There is much confusion about the degree of hydration of the chloride ion. The hydration number is generally assumed to be ≤ 6 (POWELL *et al.* 1988). IMPEY *et al.* (1983) define a dynamic hydration number which is 2.6 at 287 K. MARCHESE & BEVERIDGE (1984) suggest a nine-fold hydration. SAMOILOW (1957) assumes a negative hydration on the basis of calculations, which show that an H_2O molecule stays next to a chloride ion for a shorter period than next to another H_2O molecule. On the basis of a hydration number of 6, the diffusion coefficient ratio would be 1.0080.

Secondly, the isotopic fractionation factor can be determined experimentally from mobility ratios in the solution. Experimentally determined mobility ratios were reported to be 1.00115 to 1.00207 in a 1:30 NaCl solution (about 35 g/l, MADORSKY & STRAUSS 1947), and 1.0009 to 1.0011 and 1.0013 to 1.0015 in aqueous solutions containing 1 and 5.4 g NaCl/l respectively (KONSTANTINOV & BAKULIN 1965). It should be realized that these ratios were determined before significant improvements in analytical techniques were made, and that part of the differences may perhaps be related to differences in the electrolyte solution.

Thirdly, the chlorine isotope fractionation factor due to diffusion can be obtained by modeling $\delta^{37}Cl$ versus depth profiles in pore water systems that are diffusion controlled. Recently, DESAULNIERS *et al.* (1986) reported a decrease in dissolved chlorine content and $^{37}Cl/^{35}Cl$ ratios in Holocene sediments with increasing distance from a saline bedrock. The decrease of $^{37}Cl/^{35}Cl$ with distance was attributed to the higher diffusion coefficient of ^{35}Cl relative to ^{37}Cl. An isotope fractionation factor due to diffusion of 1.0012 was required to successfully describe the $\delta^{37}Cl$ profile with a mathematical model. A sensitivity analysis indicated that the model was very sensitive to the isotope fractionation factor, the water advection rate, and the diffusion time.

In this study we will follow the third approach to determine the chlorine isotope fractionation factor. We studied three cores from Kau Bay, island of Halmahera (Indonesia), for which the Quaternary history has been well documented (MIDDELBURG 1991; MIDDELBURG & DE LANGE 1989; MIDDELBURG *et al.* 1990, 1991). High-precision $\delta^{37}Cl$ measurements were made on pore-water samples and the chlorine isotope fractionation factor was determined by fitting the data to a diffusion model.

KAU BAY

Kau Bay is enclosed by the two northern arms of the island Halmahera (northern Maluku, Indonesia). It covers an area of 60 by 30 km (Fig. 1) and has a maximum depth of about 470 meters. The bay is separated from the Pacific Ocean by a flat-floored 30-km-wide sill, which is only 40 m below sea level. At present, Kau Bay is a semi-euxinic marine basin where inputs of low oxygen, non-sulphidic bottom water alternate with periods of anoxic, sulphidic bottom water (MIDDELBURG 1991). During the last glacial period, the sea level dropped below the sill and Kau Bay was isolated, resulting in the deposition of fresh- and brackish-water sediments (MIDDELBURG & DE LANGE 1989; MIDDELBURG *et al.* 1991). These sediments are characterized by the absence of marine microfossils, by extremely low percentages of mangrove pollen, and by the presence of a fresh- and brackish-water diatom assemblage (BARMAWIDJAJA *et al.* 1989). After the glaciation, the sea-level rose and saline water entered the bay again. Since then saline water has diffused downward into the fresh- and brackish-water sediments deposited during the glacial period. As a consequence, pore-water chloride concentration decreases with depth (MIDDELBURG & DE LANGE 1989).

Fig. 1: Map of Kau Bay showing the positions of the cores used in this study (dots) and the position of the hydrographic cast (cross).

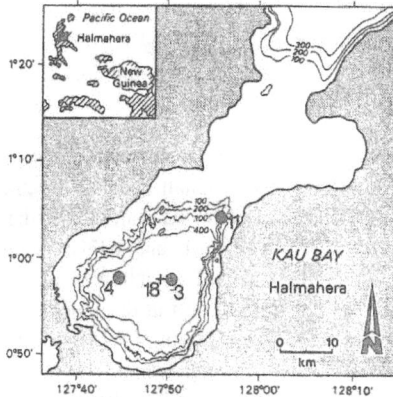

MATERIAL AND METHODS

In this study results of the analysis of three cores recovered during the Snellius II expedition in April 1985 (FIG. 1) will be discussed. The shipboard routine has been described in detail elsewhere (DE LANGE 1992). Four samples were measured from station K3, twelve from K4 and 15 from K11. Also three samples of present day bay water (hydrographic cast № 18) were collected in Niskin bottles at different depths (5, 100 and 470 meter, VAN DER WEIJDEN *et al.* 1989)

Pore water was extracted by pressure filtration of the sediment in Reeburgh-type squeezers in a nitrogen-filled glove-box. Dissolved chloride was analyzed by potentiometric titration using $AgNO_3$. The results have a relative standard deviation of less than 0.1 %.

Our method for the determination of $\delta^{37}Cl$ is an improvement of the method that was originally developed by TAYLOR & GRIMSRUD (1969) and modified by KAUFMANN (1984). About 0.1 mmol Cl⁻ is required for each $\delta^{37}Cl$ determination, i.e. 185 to 250 µl of samples having chloride concentrations that vary between 400 and 540 mmol.kg⁻¹ (14000 and 19000 ppm). To produce a constant ionic strength and a constant pH, 4 ml 1N KNO_3 and 2 ml citric acid/phosphate buffer are added to the sample. This mixture is heated to approximately 100 °C and 1 ml of 0.2N $AgNO_3$ is added. The resulting AgCl precipitate is filtered on a Whatman® glass fibre filter (GF/F, retention 0.7 µm). The filter is dried over-night in an oven at 80 °C and the AgCl yield is determined by weighing. The filter with AgCl and 200 µl CH_3I are sealed together in an evacuated pyrex® vacuum tube and reacted for two days at 80 °C. Temperatures higher than 80 °C result in partial decomposition of the CH_3I, and the $\delta^{37}Cl$ measurements become less reproducible. In the sealed tube AgCl reacts with CH_3I to form AgI and CH_3Cl. After breaking the tube under vacuum CH_3Cl and the remaining CH_3I are separated on a gas chromatograph. The isotopic composition of CH_3Cl is measured with a SIRA 24 mass spectrometer with adjustable collectors. With this method the standard deviation of a series of measurements (including sample preparation) is generally less than 0.1‰.

RESULTS

FIG. 2 and **table 1** show dissolved chloride concentrations and $\delta^{37}Cl$ as a function of depth. At station K3, there is only a very small decrease in chloride concentration and δ37Cl values show no significant variation. At station K4, the chloride concentration decreases with depth from 540 to 477 mM, and δ37Cl values range from -0.14 to +0.06‰, which is about twice the standard deviation of our method. At station K11, chloride concentration decreases from 540 mM at the sediment-water interface to about 415 mM at a depth of 7 meter. δ37Cl ranges from +0.07 to -0.31‰.

Regression analysis of δ37Cl values versus depth (**table 2**) indicates that the δ37Cl decrease with depth is not significant at station K3, slightly significant at station K4 and significant at station K11. The intercepts of the regression of δ37Cl are about +0.05‰ for the stations K4 and K11. This is a few hundredths of a per mill higher than standard sea water (SMOC), but lower than δ37Cl of present day Kau Bay water which varies from 0.16 to 0.22‰.

Table 1: Depth, chloride concentration and δ37Cl in cores K3, K4 and K11 and cast 18. Depth in cast 18 is the depth below sea-level. In K3, K4 and K11 it is the depth below the sediment-water interface.

Depth (m)	Chloride conc. (mM)	δ^{37}Cl (‰vs. S.M.O.C.)
Core K3		
0.02	539	-0.07±0.05
2.71	541	-0.03±0.04
5.51	537	-0.02±0.03
7.03	530	-0.08±0.02
Core K4		
0.12	539	0.03±0.01
1.46	534	0.06±0.06
3.15	519	0.05±0.06
4.15	507	-0.08±0.10
5.08	501	0.01±0.06
5.55	496	-0.02±0.12
6.11	485	-0.03±0.03
6.43	482	-0.14±0.06
6.89	490	-0.01±0.08
7.08	484	0.02±0.00
7.25	484	0.00±0.05
7.56	477	-0.10±0.06
Core K11		
0.05	539	0.00±0.16
0.40	534	0.00±0.00
0.75	530	0.06±0.18
1.45	514	0.07±0.18
1.78	504	-0.09±0.10
2.43	487	-0.11±0.01
2.76	477	-0.06±0.04
3.14	473	-0.06±0.04
3.36	470	-0.10±0.04
3.63	461	-0.18±0.03
4.46	452	-0.22±0.10
4.68	445	-0.21±0.06
5.64	434	-0.30±0.10
6.17	422	-0.31±0.11
6.71	415	-0.27±0.08
Bay water (Cast 18)		
5	-	0.16±0.02
100	-	0.22±0.05
470	-	0.19±0.02

DISCUSSION

In order to successfully model δ37Cl values and hence to constrain the chlorine isotope fractionation factor, it is mandatory to develop a model which can describe both the dissolved chlorine concentrations and the δ^{37}Cl values.

To this extent the model of MIDDELBURG & DE LANGE (1989) was modified to calculate δ37Cl values. Assuming that chlorine is not involved in chemical reactions and that there are no sorption effects, the differential equation becomes:

$$\frac{\partial C}{\partial t} = D\left(\frac{\partial^2 C}{\partial x^2}\right) - V\left(\frac{\partial C}{\partial x}\right) \tag{3}$$

Table 2: Regression parameters for K3, K4 and K11. Independent variable is depth, dependent variable is δ37Cl.
Regression parameters:

	number of observations	line (intercept and slope)	Correlation coefficient (r)
Core K3	4	δ^{37}Cl=-0.051-0.000D	-0.03
Core K4	12	δ^{37}Cl=0.050+0.013D	+0.50
Core K11	15	δ^{37}Cl=0.057+0.056D	+0.93

where D is the diffusion coefficient, C is the concentration of Cl⁻, x is the depth below the sediment-water interface, t is the time from the beginning of diffusion and V is the pore water advection rate. The appropriate initial upper and lower boundary conditions are: at t=0 the chloride concentration in the pore water is constant with depth, whereas at t>0 the chloride concentration at the water-sediment interface is equal to the bay water chloride concentration and as t>0 and depth approach infinity, there is no gradient $\partial C/\partial x$=0. The solution of equation 3 for these initial and boundary conditions is:

$$C = C_i + (C_0 - C_i)\, A_{(x,t)}$$
$$\text{where}$$
$$A_{(x,t)} = \frac{1}{2}\, erfc\,\frac{(x-Vt)}{2\sqrt{Dt}} + \frac{1}{2}\exp\left(\frac{Vx}{D}\right)erfc\,\frac{(x+Vt)}{2\sqrt{Dt}} \tag{4}$$

where C_i is the initial concentration of the chlorine in fresh water and C_0 is the concentration in seawater and Erfc is the error function complement (CARSLAW & JAEGER 1959, CRANK 1975).

Five variables are required to describe present-day chloride concentration versus depth profiles, namely C_o, D, V, t and C_i. The chloride concentration of the bottom water (C_o) is 540 mM at all three sites. The porosity of Kau Bay sediments is surprisingly constant at 0.81, with no significant changes with either depth or location (MIDDELBURG & DE LANGE 1989) Based on this porosity and appropriate estimates of the temperature corrected free diffusion coefficient (LI & GREGORY 1974) and formation factor (ULLMAN &

Fig. 2: Dissolved chloride concentration and $\delta^{37}Cl$ in core K3, K4 and K11 as a function of depth. The regression line of $\delta^{37}Cl$ as a function of depth is also shown.

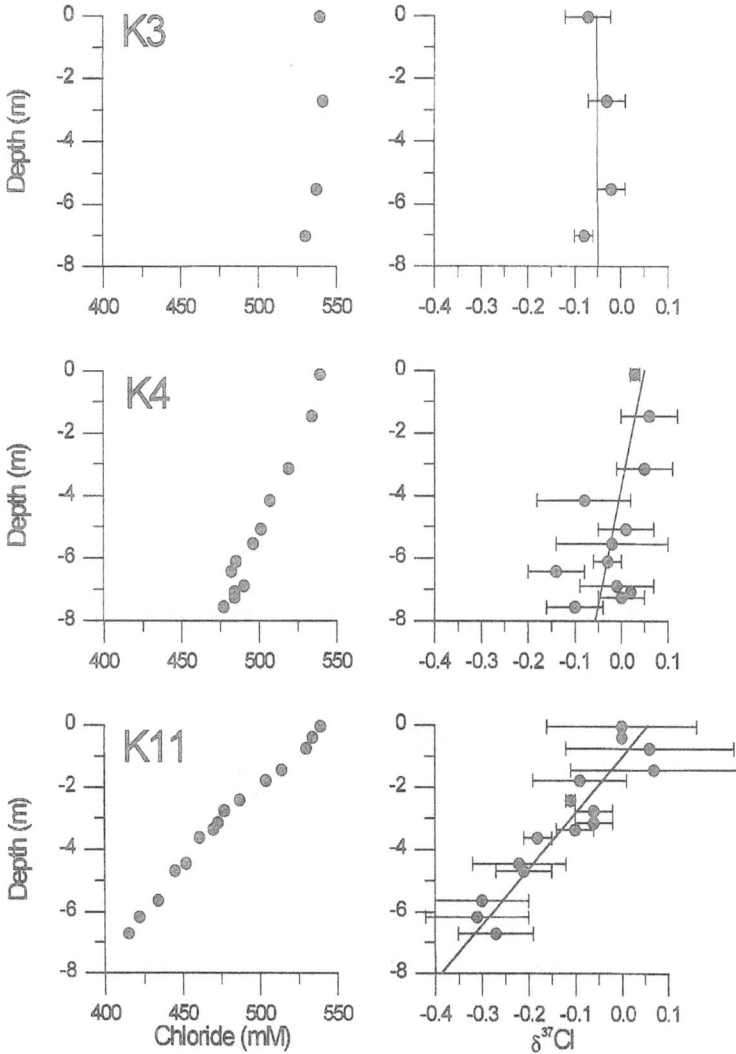

ALLER 1983), a sediment diffusion coefficient of 413 cm²yr⁻¹ (13.1 m²s⁻¹) is obtained. There are two constraints on the time (*t*) estimate. Firstly, at station K4 the transition from fresh and brackish to marine sediments occurs at a depth of about 750 cm. The sediment accumulation rate at this site is 71 cm/kyr (MIDDELBURG 1991). Accordingly, the transition took place about 10000 years ago. This date is supported on the basis of a sill depth of 40 m and the established sea-level curve (e.g. see BLOOM *et al.* 1974), which also indicates that reconnection with the sea occurred about 10000 years ago. The sediment accumulation rate at Kau Bay has been determined by Accelerator Mass Spectrometry [14]C

Fig. 3: Determination of C_i in core K4. For line A, C_i is 125 mM, for line B it is 150 mM, and for line C it is 175 mM

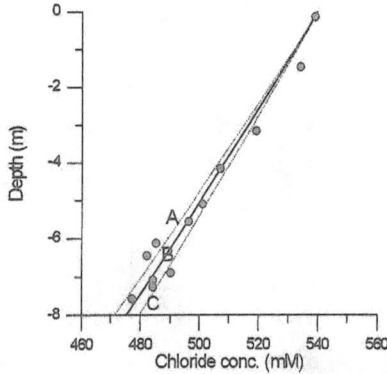

Fig. 4: Determination of the net advective rate in core K11. For line A, the net advective rate is -0.200 cm/yr, for line B it is -0.175 cm/yr, and for line C it is -0.150 cm/yr.

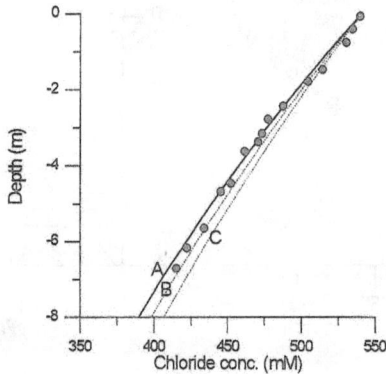

dating (VAN DER BORG *et al.* 1987) of pteropod shells and yields values of 0.6, 0.071 and 0.085 cm/yr at station K3, K4 and K11 respectively. The chloride concentration of the interstitial water at the start of the diffusion process (C_i) is not known, but should be similar at all sites.

Our modeling strategy is as follows: C_o, D and t are fixed at the values given above, the pore water advection rate is taken equal to the sediment accumulation rate, and C_i is obtained by curve fitting on site K4 (FIG. 3). We were able to describe chloride

Table 3: *Variables used to draw diffuson lines in FIGS. 5 and 6.*

Core	Diffusion time (years)	Diffusion coefficient (cm²/yr)	C_0 (mM)	$C_{initial}$ (mM)	Pore-water advection rate (cm/yr)	α (diffusion coefficient ratio; D_{35}/D_{37})
K4	10000	413	540	150	0.075	1.0030
K11	10000	413	540	150	-0.200	1.0023

concentration versus depth profiles at stations K3 and K4 with an initial chloride concentration of 150 mM. Since no correlation is found between $\delta^{37}Cl$ and depth at station K3, no further calculations were done on this core. The pore water chloride profile at station K11, could only be described by assuming a net advective water flow in upward direction of about -0.2 cm/yr (FIG. 4).

$\delta^{37}Cl$ versus depth profiles can be modelled if two more variables are known, namely the initial $^{37}Cl/^{35}Cl$, ratio and the isotope fractionation factor due to diffusion. The initial $\delta^{37}Cl$ value of the pore water is assumed to be zero. This is a fair assumption given the limited fractionation of Cl in nature. The isotopic fractionation factor of chlorine due to diffusion has been determined by fitting model curves to the measured $\delta^{37}Cl$ versus depth profiles. The input parameters are given in **table 3** and the resulting fractionation factor varies from 1.0023 at station K11 (FIG. 5) to 1.003 at station K4 (FIG. 6). These results are

Fig. 5: Best fit diffusion line for core K11 (data in table 3).

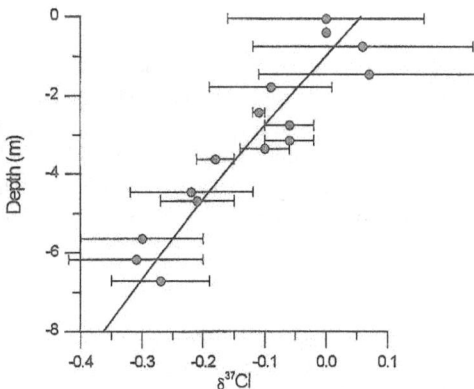

Fig 6: Best fit diffusion line for core K4 (data in table 3).

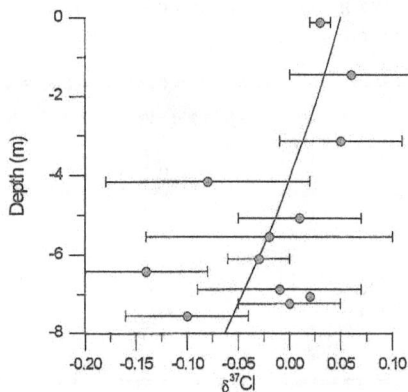

not significantly different from each other because the relative errors for K4 are about twice as large as for K11. Therefore, the value obtained for K11 is assumed to be the best approximation of the real value.

A sensitivity analysis indicated that the model predictions are primarily dependent on the advection rate and the isotopic fractionation factor, and to a lesser extent on the time and initial composition of the pore water (see also MIDDELBURG & DE LANGE 1989). In fact the modeled results are rather robust: fitting the chloride concentration and $\delta^{37}Cl$ profiles with an initial chloride concentration of zero and variable water flow rates (not corresponding to the actual sediment accumulation rates) gives almost identical isotope fractionation factors. Moreover, there is no independent evidence for any significant advection flow exept at station K11 (MIDDELBURG 1990).

CONCLUSIONS

To conclude, the chlorine isotope fractionation factor due to diffusion is approximately 1.0023 in brackish/marine sediments. This value, obtained by modelling the distribution of chlorine isotopes in a natural diffusion system agrees well with the value determined experimentally from the difference in mobility of ^{35}Cl and ^{37}Cl (1.0022; MADORSKY & STRAUSS 1947). A reliable determination of the fractionation factor is crucial to sound interpretations of chlorine isotope variations produced by diffusion.

ACKNOWLEDGMENTS

The Snellius II Expedition was an enterprise undertaken joitly by the Indonesian and the Dutch scientific community. A.F. Koster van Groos critically read an earlier version of this manuscript. D.C. McCartny is thanked for linguistic advice. This study was partly supported by the Netherlands

Foundation for Earth Science Research (AWON) with financial aid from the Netherlands Organization for the Advancement of Pure Research (NWO) (grants 751.355.012 JJM, 751.355.014 HGME). The mass-spectrometer used was partly financed by NWO.

REFERENCES

BARMAWIDJAJA D.M., DE JONG A.F.M., VAN DER BORG K., VAN DER KAARS W.A. & ZACHARIASSE W.J. (1989) Kau Bay, Halmahera, a late quaternary palaeoenvironmental record of a poorly ventilated basin. *Neth. J. Sea Res.* **24** 591-605

BLOOM A.L., BROECKER W.S., CHAPPEL J., MATTHEWS R.H. & MESOLELLA K.J. (1974) Quaternary sea level fluctuations on a tectonic coast: New $^{230}Th/^{234}U$ dates from the Huon Peninsula, New Guinea. *Quat. Res.* **4** 185-205

DE LANGE G.J. (1992) Shipboard routine and pressure-filtration system for pore-water extraction from suboxid sediments. *Mar. Geol.* **109** 77-81

CARSLAW H.S. & JAEGER J.C. (1959) *Conduction of heat in solids.* Oxford Univ. Press, London. Second Edt. 510 pp.

CRANK J.(1956) *The mathematics of diffusion.* Oxford Univ. Press, London. 347 pp.

DESAULNIERS D.E., KAUFMANN R.S., CHERRY J.A. and BENTLY H.W. (1986) ^{37}Cl-^{35}Cl variations in a diffusion-controlled groundwater system. *Geochim. Cosmochim. Acta* **50** 1757-1764

EASTOE C.J. & GUILBERT J.M. (1992) Stable chlorine isotopes in hydrothermal systems. *Geochim. Cosmochim. Acta* **56** 4247-4255

EASTOE C.J., GUILBERT J.M. & KAUFMANN R.S. (1989) Preliminary evidence for fractionation of stable chlorine isotopes in ore-forming hydrothermal systems. *Geology* **17** 285-288

EGGENKAMP H.G.M. (1994) $\delta^{37}Cl$; the geochemistry of chlorine isotopes. *Geol. Ultraj.* **??** ?? pp. Ph.D. Thesis, Utrecht University.

HOERING T.C. & PARKER P.L. (1961) The geochemistry of the stable isotopes of chlorine. *Geochim. Cosmochim. Acta* **23** 186-199

IMPEY R.W., MADDEN P.A. & McDONALD I.R. (1983) Hydration and mobility of ions in solution. *J. Phys. Chem.* **87** 5071-5083

KAUFMANN R.S. (1984) *Chlorine in ground water: Stable isotope distribution.* Ph.D. Thesis, University of Arizona. 137 pp.

KAUFMANN R.S. (1989) Equilibrium exchange models for chlorine stable isotope fractionation in high temperature environments. *Proc. WRI-6* 365-368

KAUFMANN R., LONG A., BENTLEY H. & DAVIS S. (1984) Natural chlorine isotope variations. *Nature* **309** 338-340

KAUFMANN R.S., FRAPE S.K., FRITZ P. & BENTLY H. (1987) Chlorine stable isotope composition of Canadian shield brines. *in* Saline water and gases in crystalline rocks, Editors: FRITZ P. & FRAPE S.K. *Geol. Ass. Canada spec. Pap.* **33** 89-93

KAUFMANN R.S., LONG A. & CAMPBELL D.J. (1988) Chlorine isotope distribution in formation waters, Texas and Louisiana. *AAPG bull.* **72** 839-844

KONSTANTINOV B.P. & BAKULIN E.A. (1965) Separation of chloride isotopes in aqueous solutions of lithium chloride, sodium chloride, and hydrochloric acid. *Russ. J. Phys. Chem.* **39** 315-318

LI Y.-H. & GREGORY S. (1974) Diffusion of ions in sea water and in deep-sea sediments. *Geochim. Cosmochim. Acta* **38** 703-714

MADORSKY S.L. & STRAUSS S. (1947) Concentration of isotopes of chlorine by the counter-current electromigration method. *J. Res. Nat. Bur. Stand.* **38** 185-189

MARCHESE F.T. & BEVERIDGE D.L. (1984) Pattern recognition approach to the analysis of geometrical features of solvation: application to the aqueous hydration of Li^+, Na^+, K^+, F^-, and Cl^-. *J. Amer. Chem. Soc.* **106** 3713-3720

MIDDELBURG J.B.M. (1990) Early diagenesis and authigenic mineral formation in anoxic sediments of Kau Bay, Indonesia. *Geol. Ultraj.* **71** 177 pp. Ph.D. Thesis, University of Utrecht.

MIDDELBURG J.J. (1991) Organic carbon, sulphur, and iron in recent semi-euxinic sediments of Kau Bay, Indonesia. *Geochim. Cosmochim. Acta* **55** 815-828

MIDDELBURG J.J. & DE LANGE G.J. (1989) The isolation of Kau Bay during the last glaciation: direct evidence from interstitial water chlorinity. *Neth. J. Sea Res.* **24** 615-622

MIDDELBURG J.J., DE LANGE G.J. & KREULEN R. (1990) Dolomite formation in anoxic sediments of Kau Bay, Indonesia. *Geology* **18** 399-402

MIDDELBURG J.J., CALVERT S.E. & KARLIN R. (1991) Organic rich transitional facies in silled basins: Response to sea-level change. *Geology* **19** 679-682

MORTON R.D. & CATANZARO E.J. (1954) Stable chlorine isotope abundances in apatites from Ødegårdens verk. *Norsk Geol. Tiddskr.* **44** 307-313

OWEN H.R. & SCHAEFFER O.A. (1954) The isotope abundances of chlorine from various sources. *J. Amer. Chem. Soc.* **77** 898-899

POWELL D.H., BARNES A.C., ENDERBY J.E., NEILSON G.W. & SALMON P.S. (1988) The hydration structure around chloride ions in aqueous solution. *Faraday Discuss. Chem. Soc.* **85** 137-146

SAMOILOV O.Ya. (1957) A new approach to the study of hydration of ions in aqueous solutions. *Disc. Faraday Soc.* **24** 141-146

TAYLOR J.W. & GRIMSRUD E.P. (1969) Chloride isotopic ratios by negative ion mass spectromtry. *Anal. Chem.* **41** 805-810

ULLMANN W.J. & ALLER R.C. (1982) Diffusion coefficients in nearshore marine sediments. *Limnol. Oceanogr.* **27** 552-556

VAN DER BORG K., ANDERLIESTEN C., HOUSTON C.M., DE JONG A.F.M. & VAN ZWOL N.A. (1987) Accelaration mass spectrometry with ^{14}C and ^{10}B in Utrecht. *Nuclear Instr. and Methods* **B29** 143

VAN DER WEIJDEN C.H., DE LANGE G.J., MIDDELBURG J.J., VAN DER SLOOT H.A., HOEDE D. & SHOFIYAH S. (1990) Geochemical characteristics of Kau Bay water. *Neth. J. Sea Res.* **24** 583-589

CHAPTER 6

Chlorine Isotope Ratios in
Pore Waters From the Dutch IJsselmeer Sediments;
Diffusion and Mixing

H.G.M. Eggenkamp[1], H.E. Beekman[2], C.A.J. Appelo[3] and R. Kreulen[1]

ABSTRACT-- $\delta^{37}Cl$ values are measured in pore water samples from a sediment core from the Dutch IJsselmeer. The IJsselmeer is an artificial lake in the centre of the Netherlands that was formed in 1932. Before 1932 it was a saline inlet of the North Sea, called the Zuiderzee. Before 1570 the water was brackish. On the basis that this history is probably reflected in the variations of $\delta^{37}Cl$ values of the pore water samples, these variations are examined using different diffusion models, ranging from simple analytical to advanced numerical. In the analytical models one cannot account for changes in boundary chloride concentrations, and using historical correct input parameters the chloride concentrations and $\delta^{37}Cl$ could not be modelled correctly: calculated $\delta^{37}Cl$ values were always lower than the measured values. In the numerical model it was possible to include time variations in the boundary values of chloride concentrations, and also the mixing of pore water and overlaying water during (re)sedimentation, e.g. with storms. When mixing with more saline water was implemented it was possible to succesfully model both chloride concentration and $\delta^{37}Cl$.

INTRODUCTION

The stable isotopes of chlorine fractionate only very little in nature. The main reason for the small fractionation is that chlorine normally exists only in one oxidation state. Significant fractionation, however, can occur upon diffusion, since ^{35}Cl diffuses faster than ^{37}Cl. In this study the $\delta^{37}Cl$ values of pore waters from a sediment core from the Dutch IJsselmeer are measured. The IJsselmeer has a complex geologic history, with alternating periods of fresh and saline water. We try to explain not only the observed chloride concentrations (VOLKER 1961, VOLKER & VAN DER MOLEN 1991), but also the $\delta^{37}Cl$ values using a model that takes into account this history.

-Earlier chlorine isotope studies

Fractionation of chlorine isotopes in nature is small, and the study of their behaviour in geologic systems is a young field of research. Early chlorine isotope measurements in the fifties and sixties (MORTON & CATANZARO 1954, OWEN & SHAEFFER 1954, HOERING & PARKER 1961) did not detect variations from the standard outside the limits of precision.

1Department of Geochemistry, Utrecht University, P.O.Box 80.021, 3508 TA Utrecht, The Netherlands
2Centre for Development Cooperation Services, Vrije Universiteit, De Boelelaan 1115, 1081 HV Amsterdam, The Netherlands
3Institute of Earth Sciences, Vrije Universiteit, De Boelelaan 1085, 1081 HV Amsterdam, The Netherlands

Since the early eighties, mass spectrometers have become more accurate and sample preparation procedures have been greatly improved. Since then Kaufmann and coworkers published several papers on this topic (e.g. KAUFMANN *et al.* 1984, 1987, 1988, 1992, 1993, EASTOE *et al.* 1989, EASTOE & GUILBERT 1992). The diffusion of chlorine and the related isotope effects were studied by DESAULNIERS *et al.* (1986) in ground water samples from glacial deposits in Canada. They found that during the diffusion of pore water out of a saline bedrock, the chloride concentration and the $^{37}Cl/^{35}Cl$ ratio both decrease in upward direction, indicating that the heavier ^{37}Cl isotope has a smaller diffusion coefficient than the lighter ^{35}Cl isotope.

THE IJSSELMEER

The IJsselmeer is an artificial lake in the centre of the Netherlands that was formed in 1932 by closing the then existing Zuiderzee (a brackish inlet of the North Sea in the Dutch central lowland) by a dam. Chloride concentrations in IJsselmeer sediments were examined by VOLKER (1942, unpublished: 1961) in order to calculate seepage to nearby, low-lying, reclaimed areas. In his profiles VOLKER (1961) found a sharp increase in cloride concentration in the first few maters, followed by a gradual decrease. It was shown that diffusive transport changed pore water salinity in the centuries that the lagoon was brackish/saline. The pore water was refreshed after the close of the Zuiderzee when the water became fresh. In 1987 one core was resampled especially to measure cations (BEEKMAN 1991) and to model transport of water and chemical reactions (APPELO & BEEKMAN 1992).

-Geologic history

During the Saalien boulder clay was deposited locally. As result of the ice-movement the area developed a strong relief. In the interglacial period that followed (Eemien), the relief was filled by sedimentation of marine sand and clay. During the last ice-age (Weichselien) fluviatile sands were deposited. At the end of the Pleistocene sand was blown away and dunes were formed (RIJKSDIENST VOOR DE IJSSELMEERPOLDERS 1976).

In the beginning of the Holocene a large part of the area that is now the North Sea and the Netherlands was above sea level. Although in Pleistocene times the area was flooded repeatedly (DE VRIES 1981) the sediments probably contained fresh water at the beginning of the Holocene. The sediments must have been freshened in the Weichselien. During the first 5000 years of the Holocene the western part of the Netherlands was flooded by the sea and . The originally fresh ground water became saline. After 5000 b.p. peat started to grow behind coastal barriers and the parts of the area freshened again. From about 4000 b.p. the area was flooded during a transgression phase and the marine Calais deposits were sedimented. About 1200 B.C. all tidal inlets were closed and the area became fresh. Large amounts of peat (Holland Peat) were formed in this period. In Roman times most of the western part of the Netherlands was covered by peat (GIESKE 1991). In this period the geography was probably as shown in FIG. 1a. The lake in the centre, which in fact was made up of many small and shallow lakes and marshes was called Flevomeer (Mare Flivium). In this period the older deposits freshened. During the medieval transgression (800-1200 A.D) low lying peat areas in the northwest of the Netherlands, flooded (compare Figs. 1b and 1c). In this period the lake enlarged to a lagoon sea, called Almere. From about 1300 it was called the Zuiderzee. The enlargement partly was caused by human activities such as the cultivation of the land. Although the Zuiderzee had an open connection with the sea, the water was brackish. This was caused by a large discharge of fresh water by the river IJssel, a branch of the Rhine. Many storms afflicted the area (GOTTSCHALK 1971, 1975). These storms brought saline water from the North Sea into the Zuiderzee basin and mixed it with the brackish water. In about 1570 the Zuiderzee turned saline because the discharge of the IJssel was strongly reduced (WIGGERS 1955, YPMA 1962, ENTE et al. 1986). In 1932 the Zuiderzee was closed off from the sea by an artificial dam to become a freshwater lake, the IJsselmeer. Since then large parts of the area were reclaimed (FIG. 1d). In 1973 the IJsselmeer was divided in two parts by a dam between Enkhuizen and Lelystad (see FIG. 2). The part of the lake te the west of this dam is called Markermeer.

Fig. 1: Historical geography of the IJsselmeer area (after THURKOW et al. 1984).

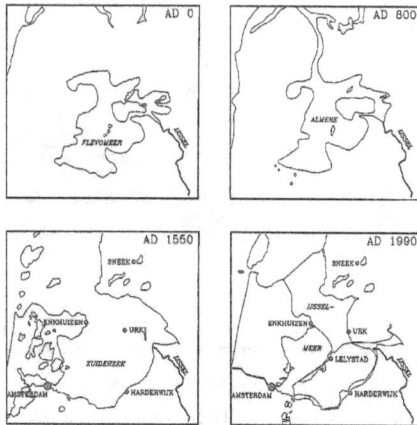

MATERIAL

-Sample location

The sediment core was collected in 1987 from aboard the motorvessel "Heffesant" of

Fig. 3: Lithostratigraphy and sediment characteristics of the sediment core (BEEKMAN 1991).

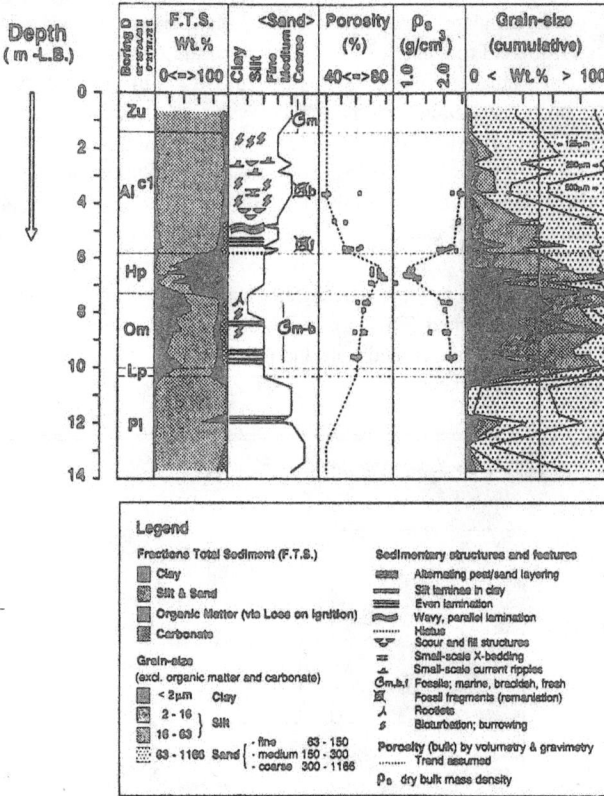

the Dutch Water Authority at location 52°39'24",48 N and 5°21'27",72 E, in what presently is the Markermeer (see FIG. 2). The bottom of the lake was 2.4 m below water surface and the core is 14 meter long. The core covers the Holocene and a few meters of Pleistocene sediments.

-Core description

The core sediments were described in detail by BEEKMAN (1991). The main stratigraphic units are shown in FIG. 3 (BEEKMAN 1991): Pleistocene (-14 to -10.3 m), Lower Peat (-10.3 to -10 m), Old Marine (Calais deposits, -10 to -7.3 m), Holland Peat (-7.3 to -5.8 m), Almere (-5.8 to -1.5 m) and Zuiderzee (-1.5 to 0 m). A hiatus exist between the Holland Peat and the Almere deposits. It is important to note that although the Old Marine deposits were formed in a marine environment, their saline pore water was replaced by fresh water during the long period that followed when the area was a fresh water lake. Therefore, the present salinity distribution reflects only the salinity changes during the Almere, Zuiderzee and IJsselmeer periods.

METHODS

Immediately after coring, the samples were frozen with liquid N_2 and stored at -20 °C during and after transportation (BEEKMAN 1991, APPELO & BEEKMAN 1992). Pore water was extracted from two 12 cm long intervals in each meter of core length. The samples were pressure filtered and analyzed after DE LANGE (1984) by G. Hamid for major cations and anions.

The method used to measure $\delta^{37}Cl$ was originally developed by TAYLOR & GRIMSRUD (1969) and was improved by KAUFMANN (1984) and ourselves. The chloride in the pore water is precipitated as silver chloride. This silver chloride is reacted with iodomethane to form chloromethane. Chloromethane and iodomethane are separated by gas-chromatography, after which chloromethane is measured in the mass spectrometer. The results are presented as ‰ deviations from a seawater sample from Madeira which is supposed to be equal to the Standard Mean Ocean Chloride isotope composition SMOC. The standard deviation of a series of measurements is smaller than 0.1‰.

RESULTS

We measured the $\delta^{37}Cl$ values of 22 pore water samples. In addition, 5 sediment samples were extracted with distilled water and also measured for $\delta^{37}Cl$. The squeezed pore water samples have numbers that end with T or B, indicating that the sample is from the top or the bottom of the indicated meter. The squeezed sediments have TB added to the sample number to indicate that the sample is between the T and the B sample (see **table 1**). Furthermore, a recent water sample from the river IJssel near the town Deventer and a sample of bottom water at the location of the sediment core was measured.

The chloride concentration data are plotted in FIG. 4, 6 and 8 and the chloride isotope data in FIG. 5, 7 and 9. Going upwards from the bottom of the sediment core a regular increase in Cl⁻ concentration from about 70 to 285mM (2500 to 10100 ppm) is manifest. This is followed by a sharp decrease in Cl⁻ concentration in the upper 2.5 meters of the core. The chlorine isotopes also show a particular pattern. $\delta^{37}Cl$ values are negative in the lower halve of the core and positive in the upper halve. From bottom to top, the $\delta^{37}Cl$ values regularly increase towards a maximum that is positioned slightly above the maximum Cl⁻ concentration, after which they sharply decrease in the upper 2.5 meters of the core.

The $\delta^{37}Cl$ values of the extracted samples generally agree with those of the pore water analyses (possibly with the exception of the low value of 14TB). $\delta^{37}Cl$ of the present river IJssel water and the IJsselmeer bottom water are close to 0‰ as expected.

Table 1: *Sample number, depth below lake bottom, chloride concentration and $\delta^{37}Cl$ of the pore water samples.*

Sample	Depth (m)	Cl⁻ (mM)	$\delta^{37}Cl$ (‰)
M1T	0.45	27	0.47±0.03
M1B	0.79	35	0.39±0.16
M2B	1.72	55	0.59±0.10
I3T	2.37	203	0.89±0.15
I3TB	2.56	-	0.61±0.10
I3B	2.74	272	0.66±0.12
I4T	3.40	273	0.53±0.07
I4TB	3.54	-	0.55±0.09
I4B	3.77	285	0.44±0.10
I5T	4.50	261	0.50±0.05
I5B	4.74	240	0.37±0.19
I6T	5.42	229	0.20±0.09
I6B	5.79	214	0.27±0.07
I7T	6.49	190	-0.01±0.10
I7B	6.79	184	-0.03±0.11
I8T	7.37	158	-0.06±0.05
I8TB	7.51	-	-0.02±0.06
I8B	7.79	146	-0.20±0.10
I9T	8.36	135	-0.30±0.00
I9B	8.79	129	-0.15±0.15
I10T	9.41	112	-0.33±0.05
I10TB	9.74	-	-0.47±0.02
I11T	10.39	91	-0.18±0.10
I12T	11.39	81	-0.45±0.10
I13T	12.51	94	-0.45±0.08
I14TB	13.45	-	-0.61±0.07
I14T	13.59	72	-0.20±0.07
Bottom	-	-	-0.03±0.10
River IJssel	-	-	-0.06±0.05

FITTING THE DATA IN A DIFFUSION MODEL

Both the chloride concentration and the chlorine isotope variations show a characteristic pattern suggesting that diffusion played an important role. Because the analytical data display such a simple pattern, it was supposed that perhaps a simple analytical diffusion model might suffice to describe the situation. This however proved not to be the case, and a numerical model was constructed. In the following the attempts are reported of fitting the seemingly simple data-set in various diffusion models. The models are, respectively, diffusion from a constant source (analytical, see chapter 4), diffusion with a constant sedimentation rate (analytical, see chapter 4), and a numerical model that can account for varying time periods of diffusion with corresponding input Cl⁻ concentrations and sedimentation and intermediate mixing (BEEKMAN 1991).

-Diffusion from a constant source?

From the shape of the chloride concentration curve (FIG. 4) it is concluded that, before diffusion occurred, the original chloride concentration in the lower (fresh-water) part of the sedimentary column was 70 mM (2500 ppm) and the original chloride concentration in the upper (saline) part of the column was 350 mM (12400 ppm). The original boundary between the fresh water and the saline water is assumed at 1.5 meter below the present lake bottom, where the boundary between Zuiderzee and Almere deposits is found. Diffusion time is, on historical grounds, taken to be 417 years, from 1570 (when the Zuiderzee became saline) until 1987 (when the core was drilled).

The diffusion coefficient in the sediment can be calculated as:

$$D_s = \frac{D_0}{\phi F} \tag{1}$$

Where D_0 is the free solution diffusion coefficient (LI & GREGORY 1974), D_s is the diffusion coefficient in the sediment, ϕ is the porosity and F is the formation resistivity factor. D_0 is taken from LI & GREGORY (1974) as $14*10^{-10}$ m²s⁻¹ at 10°C. According to BEEKMAN (1991) ϕ has an average value of about 0.6. F is defined by ARCHIE (1942) as:

$$F = 1/\phi^m \tag{2}$$

where the factor m is determined empirically (see e.g. ULLMANN & ALLER 1982). Generally m is found to be between 2 and 3, the lower values for porosities lower than 0.7 and the higher values for porosities higher than 0.7. In the present case the porosity is about 0.6 and m is chosen to be 2. The diffusion coefficient is then $8.4*10^{-10}$ m²s⁻¹. In FIG. 4 the modeled and the measured chloride concentrations are compared. In the lower part of the core where diffusion is in downward direction the measured values are higher than those predicted by the model. This might indicate that the chloride concentration of the Almere pore water increased above 70 mM before the diffusion of saline Zuiderzee water started. However, in this simple model variations of boundary concentrations cannot be modelled.

Using the same assumptions but a diffusion time of 55 years, a diffusion profile was also calculated for the upper system where chloride diffused upwards. The diffusion boundary here is taken at the recent sediment-water interface, since no sedimentation has occurred since the formation of the IJsselmeer. As can be seen in Fɪɢ. 4, the modeled and the measured values do not agree with each other. This is inherent to the model because calculations are done with fixed boundary conditions, while the chloride amount is not infinite.

Using the same constant source diffusion model, the isotope effects related to chloride diffusion can be calculated. $\delta^{37}Cl$ was calculated using a diffusion coefficient ratio of the two isotopes of 1.0023 (as determined in chapter 5), and a $\delta^{37}Cl$ prior to diffusion of 0‰. Fɪɢ. 5 shows that the calculated $\delta^{37}Cl$ curve does not agree at al with the measured values.

Left: Fig. 4: Model calculations for Cl⁻ according to the constant source model
Right: Fig. 5: Model calculations for $\delta^{37}Cl$ according to the constant source model.

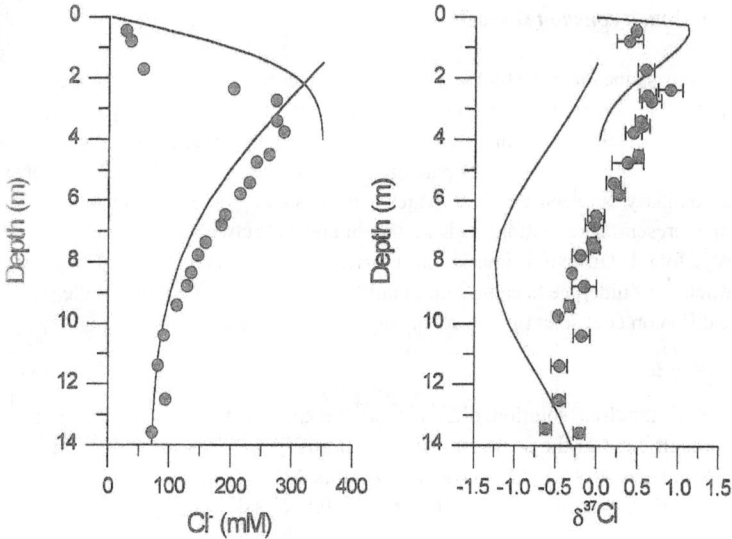

-Diffusion with a constant sedimentation rate?

While diffusion takes place, new sediment is continuously deposited on top of the older sediment, and its pore water becomes part of the diffusion system. Therefore, the actual situation will be better described by a diffusion-with-constant-sedimentation-rate model. The model used is described in the previous two chapters.

The deeper part of the column was modelled using a diffusion time of 737 years, which is in agreement with the supposed age of the sediments. At about 1250 AD the Almere became a brackish inlet after some storm surges. Since in this model the chloride concentration can not be varied, it is assumed that the concentration of 350 mM was reached instantaneously at that time. Concentration and isotope profiles are calculated both with and without advective flow of pore water. According to Volker & van der Molen (1991), a downward flow existed, at least since in 1638 polders were reclaimed and water was pumped continuously. It is assumed that all the sediment above the Holland Peat was deposited at a constant sedimentation rate of 0.79 cm.y^{-1}. The

Left: Fig. 6: Model calculations for Cl$^-$ according to the constant sedimentation model.
Right: Fig. 7: Model calculations for $\delta^{37}Cl$ according to the constant sedimentation model

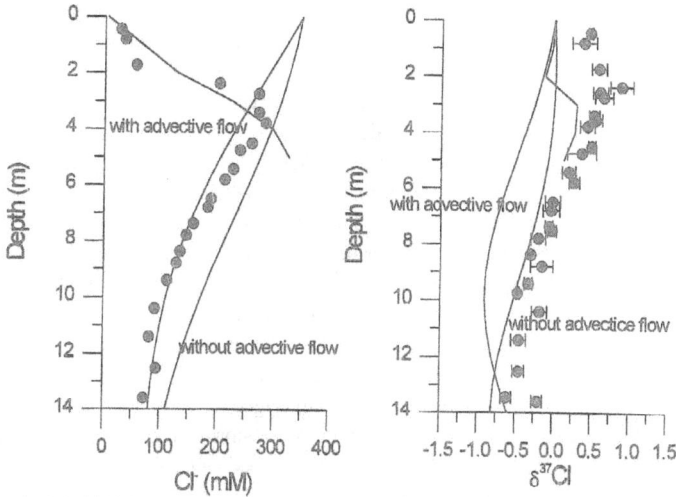

calculated chloride concentration (see Fig. 6) lies above the measured values, which may in part be due to the fact that the chloride concentration was not yet 350 mM in 1250, but increased gradually. It was tried to get a better fit by allowing for an advective flow of ground water. In Fig. 6 the curve is drawn for a net advective flow of 0.15 cm.y^{-1} (indicating an upward advective flow of 0.64 cm^{-1} and a sedimentation rate of 0.79 cm.y^{-1}). Although the fit is better, there is a total disagreement between the upward flow used in the calculations and the probability that a downward flow occurred since the reclamation of the polders.

Reversed diffusion in the upper meters of the core was calculated with only advective flow since no sedimentation occurred since the formation of the IJsselmeer. The diffusion time was 55 years (from 1932 to 1987) and the best fit is obtained when an extremely high downward advective flow of 2.7 cm.y^{-1} is assumed.

The isotope effects that can be expected in a diffusion-with-constant-sedimentation-rate model were calculated and are shown in Fig. 7. The predicted $\delta^{37}Cl$ values do not agree with the measured values; the calculated $\delta^{37}Cl$ values are between 0 to 0.6‰ too low. The agreement between calculated and measured isotopic compositions is better than in the constant-source diffusion model if no upward advective flow is assumed.

In the upper part of the profile slightly positive $\delta^{37}Cl$ values are calculated, but these are by nearly not positive enough to reach the measured values.

-Problems with the analytical models.

The results of these first two attempts to fit the measured data in a diffusion model show that it is not possible to reproduce the chloride concentrations and the chlorine isotopes, with the considered analytical models, at least not with realistic input values. This is, at least partly, due to the fact that during back-diffusion a constant source is assumed. Chloride concentrations are sometimes too high, and sometimes too low. A realistic model must also be able to vary the chloride concentration during sedimentation. Also, in all cases the calculated $\delta^{37}Cl$ values are lower than the measured ones; therefore a mechanism must be found that increases $\delta^{37}Cl$. $\delta^{37}Cl$ values increase when chloride diffuses out of the system. This is because ^{35}Cl diffuses quicker than ^{37}Cl, so that the $^{37}Cl/^{35}Cl$ ratio of the residue increases. This situation may have occurred during the Almere period after periods of heavy storm. The regular Almere water was brackish, but during storm surges saline water from the North sea was driven into the Almere. Because of the heavy wave activity the upper part of the sediments was mixed with the saline water. After the storm surge the Almere water became brackish again and excess chloride diffused out of the upper sediment layer into the overlying water and the underlying sediments. Storm surges can therefore be envisaged as momentary chloride inputs followed by chloride release. It is difficult to fit storm surges in simple analytical diffusion models, therefore a numeric model (BEEKMAN 1991, BEEKMAN *et al.* 1992) was developed. In this model it was also possible to accommodate the isotope data.

-A numerical model to solve the problem.

Three different scenarios were modeled, according to BEEKMAN (1991). In scenario (A) diffusion takes place from a fixed boundary layer, and no sedimentation occurs. In the second scenario (B) sedimentation takes place, but the effect of storm surges is not considered. Scenario (C) takes into account also the effects of mixing during storm surges.

model description

The model works as follows: The sedimentary column was divided into cells, each representing a depth interval $\Delta z = 0.5$ m. All the cells below the upper diffusion boundary plane (DBP) at depth z_d (in m below present lake bottom: -L.B.) were initially filled with the same chloride concentration (70 mM) with a $\delta^{37}Cl$ of 0‰. This chloride concentration corresponds to the concentration found directly below the Holocene deposits where the pore water salinity probably is influenced by ground water flow (see GIESKE 1989).

70

In scenario A, z_d was fixed at 1 m -L.B. The diffusion time of salt into the bottom sediments was 362 years (1570-1932). The chloride concentration and $\delta^{37}Cl$ at the bottom of the sedimentary column (at -14m -L.B.) was kept at its initial level for $t \geq 0$. In scenarios B and C the upper DBP moved upwards with time from 5.5 m to 1 m below

Table 2: *Model parameters used in the numerical program.*

	Scenario	Depth interval (m)	Time interval (y)	Cl⁻ concentration (mM)	Storm surges
Almere period	A	-	-	-	-
(1250-1570)	B	-5.5 → -1.5	320	100	-
	C	-5.5 → -1.5	320	100	Once every 40 years, 425 mM Cl⁻
Zuiderzee period	A	-1.0	362	350	-
(1750-1932)	B	-1.5 → -1.0	362	350	-
	C	-1.5 → -1.0	362	350	-
IJsselmeer period	A	-1.0	55	5	-
(1932-1987)	B	-1.0	55	5	-
	C	-1.0	55	5	-

Initial concentration in sediments below Almere: 70 mM
Diffusion coefficient ratio: 1.0023

present lake bottom. The period of brackish water input (before closure of the Zuiderzee in 1932) was set at 700 years. The upper DBP during the Al^{el} was chosen to be the sediment-bottom water interface. It is assumed that advective mixing in Al^{el} sediments below these interfaces only occurred during storm surges for scenario C. For scenario B it is assumed that storm surges had no influences on sediment pore water. During the Zuiderzee phase the DBP was chosen to be 1 m -L.B., whereas the pore water during the Zuiderzee phase was supposed to be mixed completely in the upper meter of the sediment (above the DBP) by the action of storms and bioturbation. Because the observed Cl⁻ concentrations are quite low in the upper meter of the column and can not be explained by diffusive transport, the depth of the DBP for the IJsselmeer phase was also set at 1 m. During the fresh water IJsselmeer phase (55 years) chloride concentration of 5 mM was used at the upper DBP.

Core inspection did not reveal any compaction of sediments above the Holland Peat, therefore the diffusion coefficient in the sediment (D_s) was assumed to be time independent. D_s is calculated as described above. The diffusion coefficient ratio used is 1.0023 as determined in chapter 5.

Sedimentation was simulated by adding new layers (Δz) on top of existing layers. The following cycle of events was designed in scenario C for each sedimentation step during the Al^{cl} period.

Brackish water diffuses from the DBL into underlying sediments.

During a storm event (higher surface water salinity) erosion of sediments and mixing of pore water with bottom water takes place over a certain depth interval (m_d); Cl concentration in the mixing zone becomes identical to the concentration in the bottom water during the storm and δ^{37}Cl is 0‰. After the storm, the wave activity and the salinity decreased while sediment is deposited; the DBP moves upwards and δ^{37}Cl at the top is 0‰.

In scenario B storms are not taken into account. Salt water simply diffuses from the upper DBL into the underlying sediments, DBL moves upward while sediment is deposited, and δ^{37}Cl at the surface is 0‰.

Left: Fig. 8: Model calculations for Cl⁻ according to the numerical model.
Right: Fig. 9: Model calculations for δ^{37}Cl according to the numerical model.

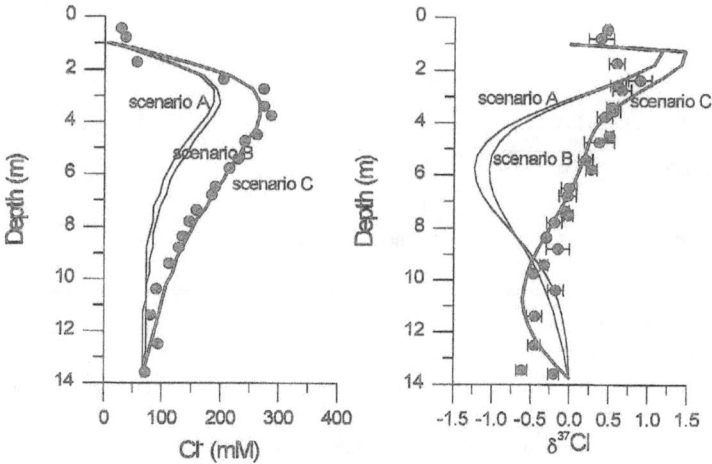

---

results and discussion

Table 2 shows the parameter values for which the figures were obtained. These values were not chosen to give the best fit, but rather represent realistic estimates for the paleoconditions. As can be seen in FIG. 8, the scenarios A and B produce too low chloride concentrations. The fit for scenario C is good. FIG. 9 shows that δ^{37}Cl fits are quite bad for the scenarios A and B, whereas scenario C gives an acceptable fit. Unfortunately, it is not exactly known what influences the storm surges had on the bottom of the lake. Based upon storm frequency (GOTTSCHALK 1971, 1975) one large storm was assumed every 40 years in the Almere period (1250-1570). During the Zuiderzee phase storm activity was much less and no mixing by storm surges is assumed. For example, in this period very large reclamation works were initiated, which would be impossible when large storm activities existed. Also GOTTSCHALK (1980) stated that in the 17th century storm surges were less than in the former centuries.

It is shown that both sedimentation and mixing/erosion is needed in the concept of the system to get acceptable results. In the models, $\delta^{37}Cl$ is set to 0‰ at a depth of 13.75m. The salinity at this depth is set to 70 mM, which is in agreement with lowest chloride concentrations in pore water from the vicinity of the core (40-85mM, e.g. GIESKE 1989). It was suggested that this chloride was derived from the underlying marine Eem deposits by upward diffusion of salt water (e.g. VOLKER 1961, THIJSSE 1972). Observations of HEBBINK & SCHULTZ (1984) of ground water flow in the first aquifer below the Holocene deposits show that the water may also come from other sources. For example ground water flow was induced by large scale reclamation activities in Northern Holland in the first half of the 17th century. It is not known at which depth this groundwater flows, but since the lowest three meters of the core consist of Pleistocene sand, it is not impossible that a lateral flow occurred in this part of the sediments. $\delta^{37}Cl$ of the pore water below the Holocene sediments is not known and it is assumed here that it is 0‰, in the absence of reliable date. We hope that in a future research these data will be available.

Differences between the three scenarios for $\delta^{37}Cl$ are the logical consequence of the modeled history of the Zuiderzee/IJsselmeer. In scenario A $\delta^{37}Cl$ is much too negative. The shape of the curve is comparable to that of the first described model (see FIG. 5). The depth where $\delta^{37}Cl$ is 0‰ is pushed down because of the additional diffusion of fresh water after 1932. (If no freshening had appeared $\delta^{37}Cl$ in the model would be 0‰ at 1m depth.) In scenario B the shape of the $\delta^{37}Cl$ curve is comparable to that in scenario A. It slightly approaches the measured values because sediments are built up gradually, with pore water of constant chloride concentration (higher than 70 mM) and $\delta^{37}Cl$ of 0‰. Because the chloride concentration was higher the isotope effect is smaller (see chapter 4) and chloride concentration was a little higher than in scenario A. However, it is apparent that the diffusion period during Almere has no significant influence. In this scenario, again the depth where $\delta^{37}Cl$ is 0‰ goes down due to back diffusion during the IJsselmeer stage. Only in scenario C have the measured $\delta^{37}Cl$ values and model calculations the same general trend. During a storm event saline water with high chlorinity mixes in the pore water and a high chlorinity pore water with $\delta^{37}Cl$ of 0‰ is formed. After the storm event the chlorinity returns to the normal lower value. Diffusion of ^{35}Cl from the saline pore water into the brackish overlying water is quicker than that of ^{37}Cl. Hence the pore water obtains a slightly positive $\delta^{37}Cl$ value. This is the reason why the depth where $\delta^{37}Cl$ is 0‰ is deeper than in the other scenarios. The $\delta^{37}Cl$ values in the lower part of the profile are higher than in the other scenarios because the difference in chlorinity between the Almere and Zuiderzee water systems is less than in the other scenarios, due to storm surches.

$\delta^{37}Cl$ in the upper part of the column was difficult to model. It is assumed that effects such as partial mixing (e.g. bioturbation), water flow, erosion and, again, storms had influenced and smoothed the diffusion profile so that $\delta^{37}Cl$ could not be examined very well. It is assumed that a much more detailed study of $\delta^{37}Cl$, including a more detailed modelling, in the upper meters of sediment pore water can solve this problem.

CONCLUSIONS

Chloride diffusion profiles Dutch IJsselmeer have been measured for $\delta^{37}Cl$. The historical conditions have a large impact on the chloride concentrations and $\delta^{37}Cl$ values. It is shown that acceptable chloride **and** $\delta^{37}Cl$ values only can be modelled with due account of several intricate aspects of sedimentation and diffusion. In systems with variable chloride concentrations and diffusion planes, it is not possible to model diffusion with a simple analytical diffusion model. In this study it is shown that determining the chlorine isotopic composition can add very important information, and, especially in the situation as described, with alternation of fresh and saline water it offers the only possibility to describe the system properly.

ACKNOWLEDGEMENTS

We like to thank the Dutch Water Authority (Rijkswaterstaat) for the collection of the core. G. Hamid is thanked for the chloride analyses. S.O. Scholten read earlier versions of this chapter and suggested various improvements.

REFERENCES

APPELO C.A.J. & BEEKMAN H.E. (1992) Hydrochemical modeling of a seawater diffusion profile, Lake Ysel, The Netherlands. *Proc. WRI* **7** 205-208

ARCHIE G.E. (1942) The electrical resistivity log as an aid in determining some reservoir characteristics. *Trans. Am. Inst. Min. Metall. Pet. Eng.* **146** 54-62

BEEKMAN H.E. (1991) *Ion chromatography of fresh- and seawater intrusion. Multicomponent dispersive and diffusive transport in groundwater.* Ph.D thesis, Free University Amsterdam. 198 pp.

BEEKMAN H.E., EGGENKAMP H.G.M. APPELO C.A.J. & KREULEN R. (1992) $^{37}Cl/^{35}Cl$ transport modeling in accumulating sediments of a former brackish lagoonal environment. *Proc. WRI* **7** 209-211

DE LANGE G.J. (1984) Shipboard pressure filtration system for interstitial water extraction. *Med. RGD* **38** 209-214

DE VRIES J.J. (1981) Fresh and salt groundwater in the dutch coastal area in relation to geomorphological evolution. *In* A.J. VAN LOON (ed.): Quaternary geology: a farewell to A.J. WIGGERS. *Geol. Mijnb.* **60** 363-368

DESAULNIERS D.E., KAUFMANN R.S., CHERRY J.A. & BENTLEY H.W. (1986) $^{37}Cl-^{35}Cl$ variations in a diffusion-controlled groundwater system. *Geochim. Cosmochim. Acta* **50** 1757-1764

EASTOE C.J., GUILBERT J.M. & KAUFMANN R.S. (1989) Preliminary evidence for fractionation of stable chlorine isotopes in ore-forming hydrothermal systems. *Geology* **17** 285-288

EASTOE C.J. & GUILBERT J.M. (1992) Stable chlorine isotopes in hydrothermal processes. *Geochim. Cosmochim. Acta* **56** 4247-4255

ENTE P.J., KONING J. & KOOPSTRA R. (1986) De bodem van Oostelijk Flevoland. *Flevobericht* **258** 181 pp.

GIESKE J.M.J. (1989) Hydrologische systeemanalyse van het IJsselmeergebied. *DGV-TNO* **OS 89-05** 52 pp.

GIESKE J.M.J. (1991) De oorsprong van het brakke grondwater in het IJsselmeergebied: diffusie, dispersie of dichtheidsstroming? *H₂O* **24** 188-193

GOTTSCHALK M.K.E. (1971) *Storm surges and river floods in The Netherlands. I. The period before 1400* Van Gorcum, Assen.

GOTTSCHALK M.K.E. (1975) *Storm surges and river floods in The Netherlands. II. The period 1400-1600* Van Gorcum, Assen.

GOTTSCHALK M.K.E. (1980) Subatlantische transgressiefasen en stormvloeden. *in Transgressies en occupatiegeschiedenis in de kustgebieden van Nederland en Belgie*. Ed. A. VERHULST & M.K.E. GOTTSCHALK 21-27

HEBBINK A.J. & SCHULZ E. (1984) Geohydrologie van het noordhollandse randgebied van de Markerwaard. *Flevobericht* **238** 60 pp.

HOERING T.C. & PARKER P.L. (1961) The geochemistry of the stable isotopes of chlorine. *Geochim. Cosmochim. Acta* **23** 186-199

KAUFMANN R.S. (1984) *Chlorine in groundwater: stable isotope distribution*. Ph.D. Thesis. University of Arizona. 137 pp.

KAUFMANN R., LONG A., BENTLEY H. & DAVIS S. (1984) Natural chlorine isotope variations. *Nature* **309** 338-340

KAUFMANN R., FRAPE S., FRITZ P. & BENTLEY H. (1987) Chlorine stable isotope composition of Canadian shield brines. *In* Saline water and gases in crystalline rocks. P. FRITZ & S.K. FRAPE (eds.): *Geol. Ass. Canada Spec. Pap.* **33** 89-93

KAUFMANN R.S., LONG A. & CAMPBELL D.J. (1988) Chlorine isotope distribution in formation waters, Texas and Louisiana. *AAPG Bull.* **72** 839-844

KAUFMANN R.S., FRAPE S.K., McNUTT R. & EASTOE C. (1992) Chlorine stable isotope distribution of Michigan Basin and Canadian Shield formation waters. *Proc. WRI* **7** 943-946

KAUFMANN R.S., FRAPE S.K., McNUTT R. & EASTOE C. (1993) Chlorine stable isotope distribution of Michigan Basin formation waters. *Appl. Geochem.* **8** 403-407

LI Y.-H. & GREGORY S. (1974) Diffusions of ions in seawater and in deep-sea sediments. *Geochim. Cosmochim. Acta* **38** 703-714

MORTON R.D. & CATANZARO E.J. (1954) Stable chlorine isotope abundances in apatites from Ødegårdens verk, Norway. *Norsk Geol. Tidsskr.* **44** 307-313

OWEN H.R. & SHEAFFER O.A. (1954) The isotope abundances of chlorine from various sources. *J. Amer. Chem. Soc.* **77** 898-899

RIJKSDIENST VOOR DE IJSSELMEERPOLDERS (1976) Markerwaard, atlas bodemgesteldheid en bodemgeschiktheid. *Flevobericht* **99** 40 pp.

TAYLOR J.W. & GRIMSRUD E.P. (1969) Chlorine isotopic ratios by negative ion mass spectrometry. *Anal. Chem.* **41** 805-810

THIJSSE J.Th. (1972) *Een halve eeuw Zuiderzeewerken 1920-1970*. H.D. Tjeenk Willink B.V. Groningen. 469 pp.

ULLMAN W.J. ALLER R.C. (1982) Diffusion coefficients in nearshore marine sediments. *Limnol. Oceanol.* **27** 552-556

VOLKER A. (1961) Source of brackish ground water in Pleistocene formations beneath the Dutch polderland. *Econ. Geol.* **56** 1045-1057

VOLKER A. & VAN DER MOLEN W.H. (1991) The influence of groundwater currents on diffusion processes in a lake bottom: an old report reviewed. *J. Hydrol.* **126** 159-169

WIGGERS A.J. (1955) De wording van het Noordoostpoldergebied. *Van Zee tot Land* **14** 215 pp.

YPMA Y.N. (1962) *De geschiedenis van de zuiderzeevisserij*. Ph.D. Thesis. University of Amsterdam. 224 pp.

CHAPTER 7

Variations of Chlorine Stable Isotopes in
Formation Waters

H.G.M. Eggenkamp[1], M.L. Coleman[2,3], J.M. Matray[4] and S.O. Scholten[5]

ABSTRACT-- $\delta^{37}Cl$ variations in west European formation waters are the largest ever measured in natural waters. Since chlorine is a conservative tracer, the $\delta^{37}Cl$ value is not much affected by chemical reactions. The correlation between chloride concentration and $\delta^{37}Cl$ was found to be negative in the Paris Basin Upper Keuper sandstone (France), whereas a positive correlation was found in the Forties Field (North Sea) and the Westland Field (The Netherlands). Values in all cases range from near zero to very negative (-1.9‰, -4.3‰ and 1.8‰ for Paris Basin, Forties and Westland respectively). Waters in the Paris Basin Upper Keuper sandstone reservoir facies represent a series of two component mixtures on basin scale dimensions. The chloride source for both components is the underlying Keuper halite, separated from the reservoir by shale. Basin-margin fluids are diluted brines. They are probably meteoric water in origin and accessed the salt via basin-margin faults. The concentrated brine of the basin centre probably was overpressured and its negative values resulted either from ultra-filtration or diffusion during cross-formational flow. The formation waters in the Forties and Westland Fields are interpreted as mixtures between saline formation water, which results from dissolution of Zechstein evaporites, and a less saline component. Diffusion from shale (probably oil-source rock) caused negative values in the dilute brine although the exact process cannot be defined yet. Aqueous fluids could have followed the same migration paths as petroleum did subsequently.

INTRODUCTION

About 20% of the volume of sediments consists of pore water (HANOR 1983). This water is referred to as formation water. The compositions of these waters vary considerably. Salinity for example ranges from near zero to about 30%. Many studies have been done on formation water; DE SITTER (1947) described how connate water could form from sea water, recognizing two different diagenetic phases. During the first phase precipitation of magnesium, calcium, sulphate and carbonate occurs, whereas during the second phase salinity gradually increases. Generally the chloride concentration increases with increasing depth (DICKEY 1969). This increase is found to range between 50 and 300 ppm/m. BREDEHOEFT *et al.* (1963) suggest that clay membranes restrict transfer of ions while water can pass through the clays. More water will be expelled from the deeper part

1Department of Geochemistry, Utrecht University, P.O.Box 80.021, 3508 TA Utrecht, The Netherlands
2BP Exploration Operating Company Ltd., BPX Technology Division, Chertsey Road, Sunbury-on-Thames, Middlesex TW16 7LN, United Kingdom
3Postgraduate Research Institute for Sedimentology, University of Reading, P.O.Box 227, Whiteknights, Reading RG6 2AB, United Kingdom
4BRGM-IMRG, Institut Mixte de Recherges Géothermiques, B.P. 6009, 45060 Orléans cedex 2, France
5Koninklijke/Shell Exploratie en Productie Laboratorium, Volmerlaan 6, 2288 GD Rijswijk (ZH), The Netherlands

of the formation, leaving the remaining formation water enriched with salt. Another type of formation water is described by LAND & PREZBINDOWSKI (1981). They suggest that Na-Ca-Cl-brines of lower Cretaceous rocks in south-central Texas are formed by a reaction of halite, detrital plagioclase and water to albite and brine. Variations in formation waters also exist between regions. For example MORTON & LAND (1987) show clear differences between four regions of formation water in the Frio Formation (Oligocene) along the Texas Gulf Coast. The four regions are characterized by 1) high salinity NaCl water, 2) low salinity NaCl water with high concentrations of organic acids, 3) Ca rich water and 4) low salinity NaCl water.

Isotope studies on formation waters showed (CLAYTON *et al.* 1966) that δD variations within a single basin are generally much smaller than δD variations between basins. However, $\delta^{18}O$ varies widely within a single basin. CLAYTON *et al.* (1966) concluded that most water is of meteoric origin, that δD was not effected, and that $\delta^{18}O$ was strongly effected by isotope exchange between water and reservoir rocks. HITCHON & FRIEDMAN (1969) used δD and $\delta^{18}O$ data of western Canadian formation waters to show that surface water was mixed with modified seawater. KHARAKA *et al.* (1973) found that oil-field brines from Kettleman North Dome have a meteoric water origin.

HOERING & PARKER (1961) found possible variations in the chlorine isotopic composition of 24 formation waters. $\delta^{37}Cl$ ranged from -0.8 to +0.6‰. However, due to the analytical precision of their analyses (±1‰) these values were inconclusive. KAUFMANN *et al.* (1988) analyzed 18 samples from Texas (Wilcox and Frio formations) and Louisiana (Weeks island). They found significant variations in $\delta^{37}Cl$ values ranging from -1.24 to +0.58‰ with an average precision of 0.12‰. A weak correlation was found between the chloride concentration and $\delta^{37}Cl$ (r^2 is 0.55 in the Texas samples). Because the data showed much scatter, interpretation of the samples was difficult. KAUFMANN *et al.* (1988) concluded that chlorine isotope ratio measurements may be indicative of the origin of formation water and that they will be useful in future studies, just because the variation is relatively large. EASTOE & GUILBERT (1992) determined $\delta^{37}Cl$ of 26 samples

Fig. 1: Approximate locations of the oil-fields in Western Europe.

from the Knox Group in Tennessee, the Gulf Coast, and Palo Duro Basins. A distinct bimodality is found for these samples, a high δ³⁷Cl group (0.0 to 0.3‰) and a low δ³⁷Cl group (-1.0 to -0.6‰). The origin of this bimodality is not known. Mixing of waters from different formations is assumed by Kᴀᴜꜰᴍᴀɴɴ *et al.* (1993) for formation waters from the Michigan Basin.

In the present study 43 samples of formation waters from different areas were measured for chlorine isotopes to detect mixing of waters with different histories on both reservoir and basin scale. Samples were collected from the Paris Basin in France, the Forties oil field in the North Sea and the Dutch Westland (Fɪɢ. 1). In addition some samples from other areas (Qatar, Norway, Papua New Guinea and Alaska) were measured.

SAMPLES AND RESULTS

-Paris basin, France

Formation water samples obtained from a collaborative research project of the European Community (Elf-Aquitaine, British Petroleum, BRGM and University of Paris VI) were analyzed for δ³⁷Cl. In this project, many parameters of both rocks and fluids are measured in different wells in order to understand the water-rock interaction processes that take place in the Paris Basin (e.g. Wᴏʀᴅᴇɴ & Cᴏʟᴇᴍᴀɴ 1992).

Samples were taken from two formations: T2, T3, T5, T6, T8, T9 and GMY1 are from the Keuper (Upper Triassic), and J1, J3 and J5 are from the Dogger (Middle Jurassic).

The centre of the Paris Basin contains oil derived from organic rich Liassic mudstones, separating both studied reservoirs. The (most important) Keuper reservoir is the Chaunoy Formation which consists of fine to coarse sandstones. The thickness of these sandstones is up to 100 meters including minor shale intercalations (Mᴀᴛʀᴀʏ *et al.*

Fig. 2: Relation between Cl⁻ and δ³⁷Cl in the Paris Basin samples. Dots are samples from the Keuper Reservoir, triangles from the Dogger reservoir.

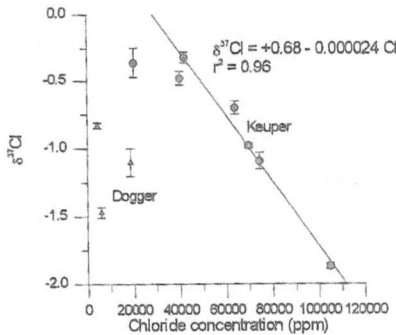

Table 1: *Most important results of analyses formation water. $\delta^{37}Cl$ values are measured in the isotope lab of Utrecht University, as are $\delta^{18}O$, δD and chloride of the Westland samples. All other analyses are done by BP exploration.*

Well	$\delta^{37}Cl$	$\delta^{18}O$	δD	Cl⁻	Br⁻	Cl⁻/Br⁻
Westland	**(Rijswijk)**					
Was27	-0.73±0.09	-3.5	-25.0	56600	n.d.	-
Was12	-0.64±0.06	n.d.	n.d.	55610	n.d.	-
Mey1	-0.71±0.08	n.d.	n.d.	56010	n.d.	-
Zoet37	-1.07±0.16	n.d.	n.d.	46690	n.d.	-
Pijn3	-0.46±0.09	-2.40	-25.0	67190	n.d.	-
Westland	**(IJsselmonde)**					
Rid13	-0.86±0.08	n.d.	n.d.	49000	n.d.	-
Westland	**(De Lier)**					
De Lier43	-1.77±0.08	n.d.	n.d.	52880	n.d.	-
De Lier23a	-1.12±0.08	-2.90	-24.0	57830	n.d.	-
Forties	**(Main Sand)**					
FA-12	-3.49±0.14	-0.85	-18.4	31500	230	137
FA-34 (1)	-2.48±0.02	-0.41	-23.5	40602	290	140
FA-34 (2)	-4.25±0.01	-1.58	-13.8	26372	165	160
FA-52	-2.00±0.04	0.34	-19.5	42800	290	148
FB-62	-1.76±0.02	0.16	-23.6	47400	n.d.	-
FC-44	-2.75±0.02	0.11	-23.8	34100	183	186
FD-43	-1.22±0.10	0.43	-25.5	48555	340	143
FD-44	-1.34±0.07	0.28	-24.5	47779	340	141
Forties	**(Charlie sand)**					
FC-23	-1.48±0.00	0.74	-13.4	32900	143	230
FC-32	-2.72±0.10	0.43	-18.0	35500	185	192
FC-41 (1)	-1.29±0.09	0.39	-4.2	29000	145	200
FC-41 (2)	-1.10±0.02	n.d.	n.d.	29320	155	189
FC-41 (3)	-1.17±0.06	n.d.	n.d.	29320	155	189
FC-41 (4)	-1.10±0.05	0.53	-13.8	28650	150	191
FC-41 (5)	-1.16±0.02	0.79	-9.1	29859	145	206
FC-41 (6)	-1.09±0.01	0.56	-11.0	32241	185	174
Paris Basin	**(Keuper)**					
T6	-1.09±0.06	-2.61	-23.1	74430	665	112
T8	-0.98±0.02	-2.71	-25.6	69832	604	116
T5	-0.70±0.05	-3.61	-20.8	63664	621	103
T9	-1.87±0.02	1.12	-12.6	104651	485	216
T2	-0.48±0.05	-4.02	-24.0	39996	365	110
T3	-0.32±0.04	-3.73	-28.5	41907	371	113
GMY-1	-0.36±0.11	n.d.	n.d.	20000	n.d.	-
Paris Basin	**(Dogger)**					
J3	-0.83±0.02	-4.43	-35.1	4284	23	186
J5	-1.10±0.10	-3.85	-18.6	18403	87	212
J3	-1.47±0.04	-4.02	-36.0	5780	40	145

Table 1, continuing.

Qatar						
DK-229	0.53±0.03	0.74	-29.9	55300	640	86
DK-229	-0.33±0.08	-2.23	-36.3	27500	280	98
DKW-1A (aq.)	0.08±0.06	-4.34	-36.8	21200	n.d.	-
DK-147	-0.26±0.07	6.70	-10.1	169900	330	515
Norway						
Gyda 2/1-6	-0.16±0.11	4.20	-31.4	159600	n.d.	-
Wildcat 2/7-22	-0.43±0.01	6.50	-23.1	90000	n.d.	-
Papua New Guinea						
PNG PPL 27	-0.57±0.08	4.08	-25.2	9100	28	325
Alaska						
Kuparuk 3F-15	-0.34±0.01	-5.76	-64.0	17100	n.d.	-
Endicott 2-14/	-0.45±0.14	-0.06	-37.9	17300	n.d.	-

1993a).

The Dogger reservoir is predominantly a carbonate assemblage with a maximum thickness of 300 meter in the centre of the basin, it consists mainly of oolitic and bioclastic limestones (MATRAY & FONTES 1990). Salinity in these fluids ranges from 1 to 32 g.l⁻¹ (MICHARD & BASTIDE 1988, MATRAY & FONTES 1990).

The seven Keuper samples show a good negative correlation ($r^2 = 0.88$, or $r^2 = 0.96$ excluding GMY1), with chloride concentrations ranging from 20000 (southern part of the reservoir, sample GMY1) to 104651 mg.l⁻¹ (eastern part which is deeper and hotter, sample T9) and δ³⁷Cl ranging from -1.87 to -0.32‰. GMY1 lies far south from the other samples and is probably not from the same reservoir (MATRAY *et al.* 1993a). In the Dogger samples no trend is recognized (FIG. 2, **table 1**).

-Forties oil field, North Sea

These waters were sampled to detect within field variations in the formation waters. Generally, it is assumed that the chemical and isotopic composition of formation water from a single well is constant. In the Forties oil field, however, it appears that during pumping both watercut and chloride concentration increased in a single well. Also, variations in chloride concentration appeared between wells in the same reservoir. These variations are probably caused by mixing of two water types, the so called Palaeo-formation water and the aquifer water. Seawater is no part of this system, since δ¹⁸O/δD variations exclude this (COLEMAN 1992).

The oil in this oil field migrated from the Kimmeridge Shale (ENGLAND 1990). The Kimmeridge Shale is the most important source rock for oils in the North Sea (e.g.

Fig. 3: Relation between Cl⁻ and δ³⁷Cl in the samples of the Forties main sand.

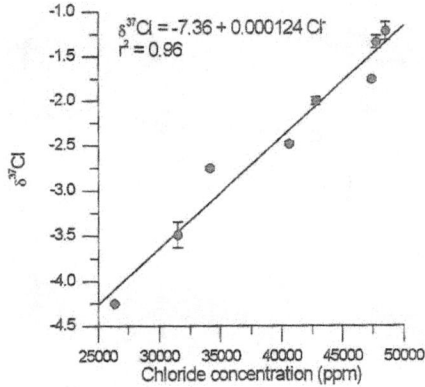

Fig. 4: Relations between Cl⁻ and δ³⁷Cl in the samples of the Forties Charlie sand. For comparison the composition of seawater is also given (triangle).

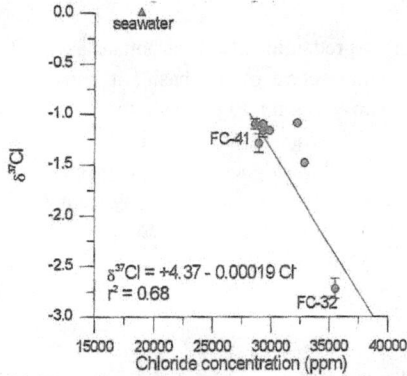

BROOKS *et al.* 1987, PETERS *et al.* 1989). The maturation of the source rock has been studied in detail (e.g. WAPLES & SLOAN 1980, MÜLLER 1977, STAHL 1977, MACKO & QUICK 1986, SCHOLTEN *et al.* 1991). This Kimmeridge source rock is effectively divided into two "kitchens", which subsided at different rates and therefore matured to different extents. The Forties reservoir is an anticlinal structure of Paleocene sands sealed with Palaeocene shales. The reservoir was filled from two sides. The bubble point pressure, for example, decreases gradually from east to west, because the more mature oil has a higher gas/oil ratio (ENGLAND 1990).

δ³⁷Cl measurements (**table 1**) were done on samples from 11 wells, some of which were sampled more than once (well FC-41, 6 times and FA-34, twice). The samples are

82

from two formations, the Main Sand (wells FA-12, -34, -52, FB-62, FC-44, FD-33, -43, and -44) and the Charlie Sand (wells FC-23, -32 and -41). The Main Sand samples are not affected by the seawater which is injected to keep pressure on the system; this is indicated by a low sulphate content in these samples (1 to 175 ppm). $\delta^{37}Cl$ in the Main Sand has a good positive correlation with the chloride content ($r^2 = 0.96$). As the chloride concentration goes up from 26372 to 48555 ppm, the $\delta^{37}Cl$ increases from -4.25 to -1.22‰ (FIG. 3).

Contrary to the Main Sand samples, the Charlie samples have a relatively high sulphate content (up to 1200 ppm), probably caused by mixing with the injected seawater. A negative correlation ($r^2 = 0.68$) exists between the chloride concentration (28650 to 35500 ppm) and $\delta^{37}Cl$ (-1.09 to -2.72‰, FIG. 4, for comparison the average seawater composition is also shown).

-Westland, the Netherlands

For the Dutch NOA-1 (Nationaal Onderzoekprogramma Aardwarmte) project, chemical and physical parameters of formation waters from the Westland (Rijswijk concession) were determined. The aim of this study was to determine what reactions take place in the highly saline waters when pressure and temperature decrease during pumping (VAN DER WEIDEN 1983). $\delta^{37}Cl$ values were determined on eight samples.

Fig. 5: Relations between Cl and $\delta^{37}Cl$ in the Westland samples. Dots are samples from the sandstone members, triangles samples from the sand-shale member.

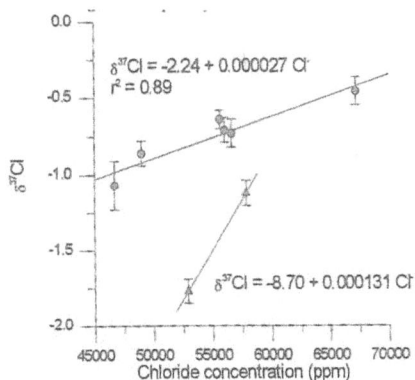

These samples, from wells in the Rijswijk concession were obtained from the Nederlandse Aardolie Maatschappij B.V. (N.A.M.) in 1983. This concession covers the mainland part of the West-Nederland Basin. From 1953 to 1988 the Rijswijk concession produced oil. Five samples were from the Rijswijk Sandstone Member (Wassenaar-27, Wassenaar-12, Meyendel-1, Zoetermeer-37 and Pijnacker-3), one from the IJsselmonde Sandstone member (Ridderkerk-13) and two from the De Lier Sand-Shale Member (De

Lier-43 and De Lier-23a). These members are part of the lower Cretaceous Vlieland Formation of middle Valanginian-Barremian age (about 125-140 Ma).

The Rijswijk Sandstone Member (Hauterivian) is a massive sandstone sequence with only a few minor shale intercalations. The IJsselmonde Sandstone Member (Barremian) is also a massive, well stratified sandstone unit with only a few clay intercalations. The De Lier Sand-Shale Member (Barremian) consists of an alternation of thin sandstones and sandy shales, commonly glauconitic, and with shell fragments and frequent bioturbation (NEDERLANDSE AARDOLIE MAATSCHAPPIJ B.V. & RIJKS GEOLOGISCHE DIENST 1980).

A good positive correlation ($r^2 = 0.89$) is found between the chloride concentration (46690 to 67190 ppm) and $\delta^{37}Cl$ (-1.07 to -0.46‰) in the Sandstone Members. $\delta^{37}Cl$ in the two Sand-Shale samples is much lower (-1.12 and -1.77‰), while the chloride concentrations of 52880 and 57830 ppm are comparable with the average value in the Sandstone samples. These samples seem to define a steeper slope (FIG. 5), but it is hard to tell with only two samples.

-Other samples

11 samples from BP oil-production fields in Qatar, Norway, Papua New Guinea and Alaska were measured. These samples are used to show the variation of $\delta^{37}Cl$ in different systems. $\delta^{37}Cl$ in these samples ranges from -0.57 to +0.08‰ and the chloride concentration from 9100 to 169900 ppm (FIG. 6).

Fig. 6: Relations between Cl and $\delta^{37}Cl$ in the other samples.

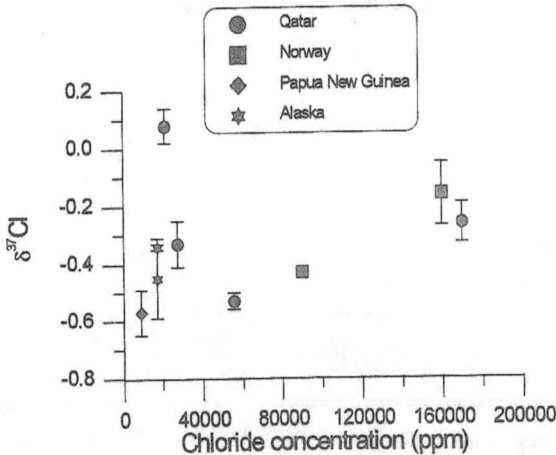

DISCUSSION

Two important correlations between chloride concentration and $\delta^{37}Cl$ were recognized: 1) a negative correlation found in the Keuper samples of the Paris Basin and 2) a positive correlation found in the Main Sand of the Forties and in the Westland. A possible negative correlation is assumed in the samples from the Charlie Sand of the Forties Basin. No correlations are found in the other samples.

Fig. 7: Relations between δ³⁷Cl and δ¹⁸O and δD in the Keuper samples of the Paris Basin.

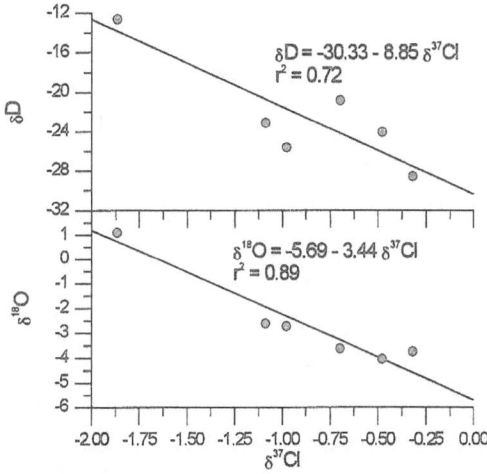

$$\delta D = -30.33 - 8.85\ \delta^{37}Cl$$
$$r^2 = 0.72$$

$$\delta^{18}O = -5.69 - 3.44\ \delta^{37}Cl$$
$$r^2 = 0.89$$

-A negative Cl⁻-δ³⁷Cl correlation in the Keuper of the Paris Basin.

For the seven Keuper samples a good negative correlation is found between the chloride concentration and $\delta^{37}Cl$ ($r^2 = 0.88$). If the southernmost sample (GMY1) is omitted, thus considering only samples from the Chaunoy reservoir, the correlation improves ($r^2 = 0.96$, FIG. 2). This high correlation is indicative of two component mixing (MATRAY *et al.* 1993a). $\delta^{18}O$ and δD values of the least saline samples are more negative than the assumed variations of seawater during the last 150 Ma. Therefore, a seawater origin of this component is excluded (FIG. 7) It is concluded that this water is meteoric water with dissolved halite.

When evaporites are formed they are depleted with respect to bromide, while the solution becomes progressively richer in bromide. Relatively high Cl/Br ratios found in the Keuper formation water samples (FIG. 8) suggest that the meteoric water dissolved a large proportion of halite with small amounts of a highly evaporated brine probably contained in fluid inclusions. This hypothesis is given by FONTES & MATRAY (1993) and

Fig. 8: Br and Cl distribution of brines from the Keuper aquifer in the Paris Basin showing the evolution path of seawater during evaporation (G means beginning of Gypsum deposition, H: Halite, E: Epsomite, S: Sylvite, C: Carnallite and B: Bischofite). Solid triangles = distribution of solid chlorides from halite (beginning = a, end = b), sylvite (c, d), carnallite (e, f) and bischofite (g). From Matray et al. (1993a).

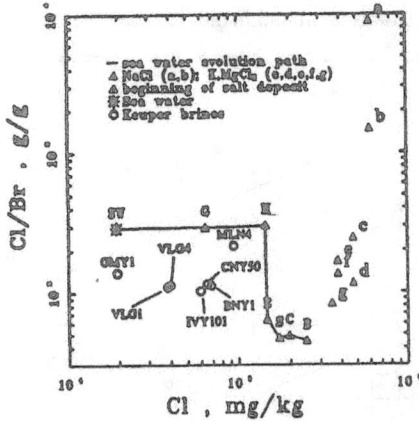

may be easily explained by dissolution of halite containing fluid inclusions of a very evolved (Br-rich) primary brine.

Ion filtration can be a possible cause for the low $\delta^{37}Cl$ values in the most saline samples (PHILLIPS & BENTLEY 1987). Ion filtration is a process that can fractionate ions as well as isotopes. Mudrock may act as semipermeable membranes and prevent or retard the passage of water and of charged species depending on the pressure and temperature conditions and of the characteristics of the water and rock involved. The residual solution of ultrafiltration is enriched in bromide with respect to chloride relative to the ultrafiltrate (KHARAKA & BERRY 1973). In addition, it appears that the residual solutions have higher $\delta^{18}O$ and δD values, while the ultrafiltrates are depleted in these isotopes (COPLEN & HANSHAW 1973). Chlorine isotope behaviour during ion filtration may be totally different from that of $\delta^{18}O$ and δD. Theoretical work predicts that the residual solution of ion filtration would be depleted in ^{37}Cl with an increase of chloride concentration (PHILLIPS & BENTLEY 1987). Un5fortunately, there are no natural observations to support this. But, because large $\delta^{37}Cl$ variations are found, an intense physical fractionation process affected the brine and although diffusion and depletion by ion-filtration are unlikely, we cannot preclude them. Therefore, it is possible that the saline end-member is the result of a membrane filtration effect on a bromide-poor secondary brine (MATRAY et al. 1993a).

86

Fig. 9: Relations between $\delta^{18}O$ and δD in the Forties Main and the Westland samples. Dots = Forties samples, filled triangles = Westland sandstone samples, open triangle = Westland sand-shale sample and square = seawater (SMOW). MWL is given for comparison.

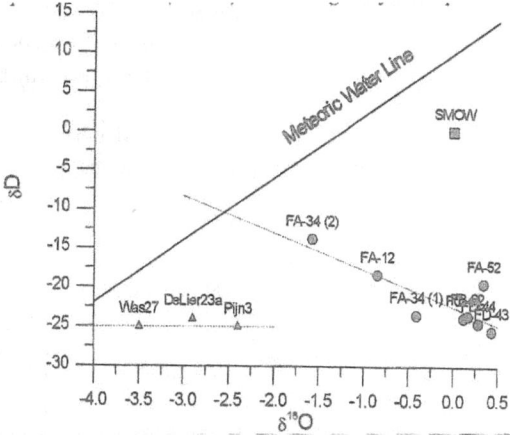

-Positive Cl^--$\delta^{37}Cl$ correlations in the Main Sand of the Forties and in the Westland.

Opposite to what was assumed before, it was discovered that the composition of formation water in the Main Sand varies within the formation, and even in individual wells (COLEMAN 1992). While a well was in production, the chloride concentration increased, together with an increase in watercut. As watercut becomes too high the well is reworked. During reworking a well, it is plugged just above the depth where water was produced and re-perforated at a higher level in the oil-production zone. Both the chloride concentration and the watercut of the well were lower after reworking. This can be explained by mixing of two different water types. Because the regression line through $\delta^{18}O$ and δD could not be extrapolated through zero, it was clear that injected seawater could not be one of these water types (see FIG. 9, COLEMAN 1992). $\delta^{18}O$ and δD suggest a possible meteoric origin of the less saline water (COLEMAN 1992). In meteoric water, however, chloride concentration is very low (near zero) and $\delta^{37}Cl$ is close to 0‰ because all chloride comes from sea water spray. The $\delta^{37}Cl$ of this particular water is extremely low (-4.25‰), so that at least the chloride can not be derived from meteoric water.

A positive correlation ($r^2 = 0.96$) is found between the chloride concentration and $\delta^{37}Cl$ (see FIG. 3). The $\delta^{37}Cl$ values near 0‰ are compatible with an evaporite source (dissolved halite) for the high chloride concentration end-member (aquifer). The origin of the low $\delta^{37}Cl$ water is more difficult to understand. A possible explanation is that water derived through dehydration of source-rock clay (smectite-illite transition) mixed with the connate water. This mixture is squeezed out of the source rock. Because of the low permeability of the source rock, this mixture is very much affected by diffusion. Since ^{35}Cl diffuses faster than ^{37}Cl, the $\delta^{37}Cl$ of this low chloride fluid will be very low (COLEMAN *et al.* 1993, EGGENKAMP & COLEMAN 1993). $\delta^{18}O$ and δD are close to the meteoric water line, not to the kaolinite line (LAWRENCE & TAYLOR 1971). This implies that the water in the original smectite had an isotopic composition close to meteoric water. In order to test this, further research on $\delta^{18}O$ and δD in these source rocks needs to be done. This end-member water may be coproduced with hydrocarbons and will be referred to as source-rock water.

The Dutch Westland samples were taken from three Members: the Rijswijk and the IJsselmonde sandstone member and the De Lier sand/shale member. A good correlation between Cl and $\delta^{37}Cl$ is found for the samples from the Sandstone members. The two samples from the sand-shale member seem to have a deviating slope in this plot (see F IG. 5). VAN DER WEIDEN (1983) assumed, based upon $\delta^{18}O$ and δD measurements that these formations waters are composed of sea and meteoric water. $\delta^{18}O$ and δD, however, seem to have a relationship that is comparable to that of the Main Sand samples (FIG. 9), thus excluding a seawater component. Because all three measured isotopes have relationships that seem to be comparable to those found in the Main Sand (see FIGS. 3, 5 and 9), it is assumed that they are formed in a similar way, thus by mixing of a source rock and aquifer water. The different slopes between the Sand- and the Sand-Shale samples are probably caused because mixing in the Sand-Shale was worse as a result of the appearance of shale that avoided mixing of the water types, resulting in a steeper $\delta^{37}Cl$-Cl slope. Unfortunately only two samples were measured.

Fig. 10: Relation between SO_4^{2-} and $\delta^{37}Cl$ in the Forties Charlie sand samples. For comparison the seawater composition is given (triangle).

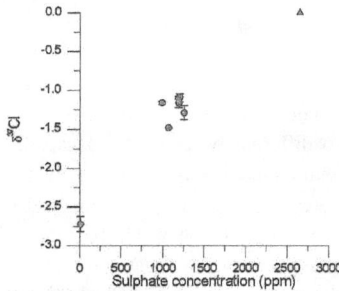

-A "negative" Cl-$\delta^{37}Cl$ correlation in Charlie Sand from the Forties Basin.

The samples from the Charlie Sand in the Forties Reservoir, have a completely different signature than the Main Sand samples. As can be seen in FIG. 4, $\delta^{37}Cl$ and chloride concentration have a negative correlation. Except for one sample (FC-32, SO_4^{2-} =

Fig. 11: Mixing line between seawater and FC-32 for Cl⁻ and δ³⁷Cl. As can be seen FC-41 is not on he mixing line.

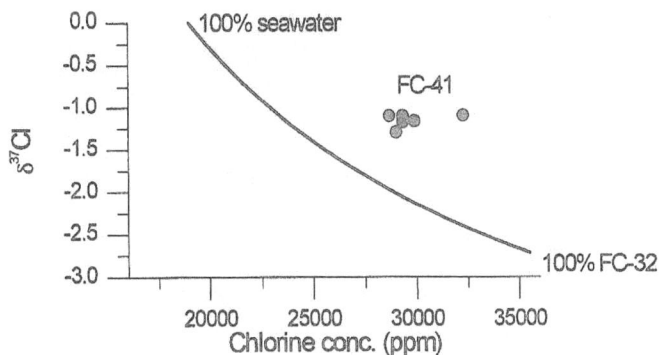

Table 2: Calculations for mixing in the Forties Charlie sand for Cl⁻, Na⁺, δD and δ¹⁸O. In these calculations two-component mixing is assumed, end-members are seawater and FC-32. Cl⁻ and Na⁺ in ppm, δD and δ¹⁸O in ‰. FC-41 min. and FC-41 max. are the lowest and highest measured values for FC-41. %sea max. and %sea min. are the lowest and highest calculated percentage of seawater in FC-41. Values with av. Are averaqge values.

	FC-32	FC-41 min	FC-41 avg	FC-41 max.	Sea	% sea min.	% sea av.	% sea max.
Cl⁻	35500	28650	29732	32241	18980	20	35	41
Na⁺	20500	16100	17308	19480	10561	10	32	44
δD	-18	-13.8	-9.5	-4.2	0	23	47	77
δ¹⁸O	0.43	0.39	0.57	0.79	0	<0	<0	9

9 ppm), the sulphate concentrations are high (990-1250 ppm). The chloride (Fɪɢ. 4) and sulphate concentrations (Fɪɢ. 10) as well as the δ³⁷Cl lie between the Main Sand and the average seawater compositions (seawater contains 2650 ppm SO_4^{2-} and 18980 ppm Cl⁻). It is therefore supposed that seawater that was injected to maintain fluid pressure during production was mixed with the original formation water. It is assumed that FC-32 represents the original formation water since its sulphate content is very low (9 ppm). Sulphate of the injected seawater has reacted with the barium of the formation water, and thus is partly removed from the solution.

A possible two component mixing calculated with Cl⁻ and Na⁺ gives for FC-41 a mixture of about 35% seawater and 65% formation water (see **table 2**). Using hydrogen isotopes indicates that the percentage seawater in sample FC-41 is higher. Calculations for possible wixing of oxygen isotopes indicate that the values for FC-41 cannot be reached by mixing the end-components, thus, the oxygen isotope composition must have been altered by a reaction. Based upon chloride concentration and chlorine isotopes it seems that a third component must be present since $\delta^{37}Cl$ values of the intermediate sample FC-41 have values that are less negative than expected from pure two component mixing (FIG. 11). The composition of this component is not clear yet, but seems to have a relatively high chloride concentration and a slightly negative $\delta^{37}Cl$.

-No correlation in Paris Basin Dogger and the other samples.

No correlations are found for the Dogger samples. However, two important points must be noted on these samples. First of all the Dogger samples display negative $\delta^{37}Cl$ values which exclude the possible conservation of a connate seawater. This hypothesis was previously proposed by MICHARD & BASTIDE (1988) to account for the similarity of TDS values between seawater and the deepest solutions from the Dogger aquifer. The study of stable isotopes of the water exclude this hypothesis (MATRAY et al. 1993b). Secondly, the $\delta^{37}Cl$ values of the Dogger samples are bracketed by those of the extreme Keuper values. This a major result which indicates that the saline load of the Dogger water is derived from input of Keuper brines into the Dogger aquifer. Such a result was also deduced from stable isotope studies of this water (MATRAY et al. 1993b)

The other samples are mostly single samples and were only measured to show that in other areas also variations in the chlorine stable isotope composition are found.

CONCLUSIONS

This study teaches us that most formation waters are mixtures of several water types. A striking result of the measurements is that $Cl^-/\delta^{37}Cl$ correlations in specific basins are all very high, and that the correlations can be positive as well as negative. The origins and compositions of the end members of these mixtures are sometimes difficult to interpret. The end member with the low $\delta^{37}Cl$, generally with extremely low values, (like the unusually low $\delta^{37}Cl$ value of the Forties oil field), is explained by either ultra-filtration or by diffusion during cross-formational flow when the chloride concentration is high, or from diffusion of water from the source rock when the chloride concentration is low. The other end member has a $\delta^{37}Cl$ that is closer to zero, suggesting a direct seawater component, or dissolution of evaporite (especially halite) deposits.

Using chlorine isotopes adds a new instrument to the description of formation waters, and possibly even to the determination of the origins of oil. Due to the lack of chemical reactions involving chlorine, that $\delta^{37}Cl$ provides one of the most conservative natural tracers for water. It is recommended that more research is done on this subject to clarify the relationships of formation water with oil.

ACKNOWLEDGEMENTS

Formation water samples are provided by NAM, BP, ESSO REP, and SNEA(P). Prof. C.H. Van der Weijden pointed us to the formation water samples from the Westland, and B. Zuurdeeg provided us with information about these samples. R. Kreulen read earlier versions of this chapter and suggested many improvements. Laboratories of BRGM, BPExploration and Utrecht University measured chloride and bromide concentrations. Part of this research (Paris Basin) is part of a collaborative project between BRGM, BP, SNEA(P), the University of Paris VI and the European Community under contract JOUF-0016-C. This research is part of the AWON project "The geochemistry of chlorine isotopes", project number 755.351.014, with financial aid from the Netherlands Organization for the Advancement of Science (NWO).

REFERENCES

BREDEHOEFT J.D., BLYTH C.R., WHITE W.A. & MAXEY G.B. (1963) Possible mechanism for concentration of brines in subsurface formations. *AAPG bull.* **47** 257-269

BROOKS J., CORNFORD C. & ARCHER R. (1987) The role of hydrocarbon source rocks in petroleum exploration. In *Marine petroleum source rocks.* (Edt. J. BROOKS & A.J. FLEET) *Geol. Soc. Spec. Publ.* **26** 17-46

CLAYTON R.N., FRIEDMAN I., GRAF D.L., MAYEDA T.K., MEENTS W.F. & SHIMP N.F. (1966) The origin of formation waters. 1. Isotopic composition. *J. Geoph. Res.* **71** 3869-3882

COLEMAN M.L. (1992) Water composition within one formation. PROC. WRI **7** 1109-1112

COLEMAN M., EGGENKAMP H.G.M., MATRAY J.M. & PALLANT M. (1993) Origins of oil-field brines by Cl stable isotopes. *Terra Abs.* **5 (1)** 638

COPLEN T.B. & HANSHAW B.B. (1973) Ultrafiltration by a compacted clay membrane. I. Oxygen and hydrogen isotopic fractionation. *Geochim. Cosmochim. Acta* **37** 2295-2310

DE SITTER (1947) Diagenesis of oil-field brines. *AAPG Bull.* **31** 2030-2040

DICKEY P.A. (1969) Increasing concentration of subsurface brines with depth. *Chem. Geol.* **4** 361-370

EASTOE C.J. & GUILBERT J.M. (1992) Stable chlorine isotopes in hydrothermal processes. *Geochim. Cosmochim. Acta* **56** 4247-4255

EGGENKAMP & COLEMAN (1993) Extreme $\delta^{37}Cl$ variations in formation water and its possible relation to the migration from source to trap. *Abstract AAPG International Conference and Exhibition.* The Hague.

ENGLAND W.A. (1990) The organic geochemistry of of petroleum reservoirs. *Org. Geochem.* **16** 415-425

FONTES J.C. & MATRAY J.M. (1993) Geochemistry and origin of formation brines from the Paris Basin. Part 2: Saline solutions associated with oil fields. *Chem. Geol. (in press)*

HANOR J.S. (1983) Fifty years of development of thought on the origin and evolution of subsurface sedimentary brines. *in: Revolution in the earth sciences. Advances in the past half-century.* Ed. S.J. BOARDMAN 99-111

HITCHON B. & FRIEDMAN I. (1969) Geochemistry and origin of formation waters in the western Canada sedimentary basin - I. Stable isotopes of hydrogen and oxygen. *Geochim. Cosmochim. Acta* **33** 1321-1349

HOERING T.C. & PARKER P.L. (1961) The geochemistry of the stable isotopes of chlorine. *Geochim. Cosmochim. Acta* **23** 186-199

KAUFMANN R.S., LONG A. & CAMPBELL D.J. (1988) Chlorine isotope distribution in formation waters, Texas and Louisiana. *AAPG Bull.* **72** 839-844

KAUFMANN R.S., FRAPE S.K., MCNUTT R. & EASTOE C. (1993) Chlorine stable isotope distribution of Michigan Basin formation waters. *Appl. Geoch.* **8** 403-407

KHARAKA Y.K. & BERRY F.A.F. (1973) Simultaneous flow of water and solutes through geological membranes. I. Experimental investigation. *Geochim. Cosmochim. Acta* **37** 2577-2603

KHARAKA Y.F., BERRY F.A.F. & FRIEDMAN I. (1973) Isotopic composition of oil-field brines from Kettleman North Dome, California, and their geologic implications. *Geochim. Cosmochim. Acta* **37** 1899-1908

LAND L.S. & PREZBINDOWSKI D.R. (1981) The origin and evolution of saline formation water, lower Cretaceous carbonates, South-Central Texas, U.S.A. *J. Hydrol.* **54** 51-74

LAWRENCE J.R. & TAYLOR H.P. (1971) Deuterium and oxygen-18 correlation: Clay minerals and hydroxides in Quaternary soils compared to meteoric waters. *Geochim. Cosmochim. Acta* **35** 993-1003

MACKO S.A. & QUICK R.S. (1986) A geochemical study of oil migration at source rock reservoirs contacts: Stable isotopes. In *Advances in Organic Geochemistry 1985* (Edt. D. LEYTHÄUSER & J. RULLKÖTTER) *Org. Geochem.* **10** 199-205

MATRAY J.M. & FONTES J.C. (1990) Origin of oil-field brines in the Paris Basin. *Geology* **18** 501-504

MATRAY J.M., COLEMAN M.L. & EGGENKAMP H.G.M. (1993a) Origin of the Keuper formation waters in the Paris Basin. in *Geofluids '93 Extended Abstracts* (Edt. J. PARNELL *et al.*) 319-322

MATRAY J.M., FONTES J.C. & LAMBERT H. (1993b) Stable isotope conservation and origin of saline waters from the Middle Jurassic aquifer of the Paris Basin, France. *Appl. Geochem. (in press)*

MICHARD G. & BASTIDE J.P. (1988) Étude géochimique de la nappe du Dogger du Bassin Parisien. *J. Volcanol. Geotherm. Res.* **35** 151-163

MORTON R.A. & LAND L.S. (1987) Regional variations in formation water chemistry, Frio formation (Oligocene), Texas Gulf Coast. *AAPG bull.* **71** 191-206

MÜLLER P.J. (1977) C/N ratios in Pacific deep-sea sediments: Effect of inorganic ammonium and organic nitrogen compounds sorbed by clays. *Geochim. Cosmochim. Acta* **44** 765-776

NEDERLANDSE AARDOLIE MAATSCHAPPIJ B.V. & RIJKS GEOLOGISCHE DIENST (1980) Stratigraphic nomenclature of the Netherlands. *Verh. Kon. Ned. Geol. Mijnb. Gen.* **32** 77 pp.

PETERS K.E., FOSCOLOS A.E., GUNTHER P.R. & SNOWDON L.R. (1989) Origin of Beatrice Oil by co-sourcing from Devonian and Middle-Jurassic source rocks, Inner Moray Firth, United Kingdom. *AAPG Bull.* **73** 454-471

PHILLIPS F.M. & BENTLEY H. W. (1987) Isotopic fractionation during ion filtration: I. Theory. *Geochim. Cosmochim. Acta* **51** 683-695

SCHOLTEN S.O., MEESTERBURRIE J.A.N. & KREULEN R. (1991) Variations in $^{15}N/^{14}N$ and $^{13}C/^{12}C$-ratios during maturation of kerogen-rich shales and related oils. In SCHOLTEN S.O. *The distribution of nitrogen isotopes in sediments.* Ph.D. Thesis, Utrecht. *Geol. Ultrai.* **81** 79-98

STAHL W.J. (1977) Carbon and nitrogen isotopes in hydrocarbon research and exploration. *Chem. Geol.* **20** 121-149

VAN DER WEIDEN M.J.J. (1983) *Een hydrogeochemisch onderzoek in verband met de winning van aardwarmte uit het West-Nederland bekken, toegespitst op het demonstratieproject Delfland.* Unpubl. MSc. Thesis., University of Utrecht. 34 pp.

WAPLES D.W. & SLOAN J.R. Carbon and nitrogen diagenesis in deep sea sediments. *Geochim. Cosmochim. Acta* **44** 1463-1470

WORDEN R.H. & COLEMAN M.L. (1992) Geochemical studies of rocks and fluids to give predictive modelling of permeability distribution in a sedimentary basin. Fourth periodic report. *Contract № JOUF-0016, funded in part by The commision of the European Communities.* 42 pp.

CHAPTER 8

Chlorine Stable Isotope Variations in Volcanic Gasses, Volcanic Springs and Crater Lakes from Indonesia

H.G.M. Eggenkamp[1] and R. Kreulen[1]

ABSTRACT-- $\delta^{37}Cl$ in volcanic spring-waters and condensates is compared with $\delta^{18}O$ and δD. Based upon $\delta^{18}O$ and δD, the samples can be divided in two groups, meteoric waters, lying along the meteoric water line, and geothermal waters with relatively positive $\delta^{18}O$ values. These groups can also be recognized with the chlorine isotope composition. Meteoric waters generally have negative $\delta^{37}Cl$ values and geothermal waters generally have positive values. Weathered rocks contribute to the chloride in meteoric waters, and volcanic gasses provide the chloride for the geothermal water samples. This might indicate that chlorine that escapes as HCl from magmas is enriched in ^{37}Cl and that rocks are thus depleted in ^{37}Cl. $\delta^{37}Cl$ of gas samples collected in Giggenbach bottles have a good inverse correlation with the chloride concentration. It is assumed that $\delta^{37}Cl$ values in these samples have no geological meaning, but are a result of fractionation of chlorine isotopes during sampling, or of analytical problems caused by the chemical composition of the samples. We therefore recommend to consider reevaluation of chemical analysis done on Giggenbach bottles.

INTRODUCTION

Volcanic processes can produce substantial fractionation of chlorine isotopes. A spectacular example is the sal ammoniac sample from the 1892 eruption of Etna volcano (Sicily, Italy), collected in 1907 by F.W. Rucler. The $\delta^{37}Cl$ of this sample; -4.88‰ (chapter 12), is the lowest value measured in any geological material. This low value can be explained by isotope fractionation during repeated sublimation and condensation of NH_4Cl. Isotope fractionation can also be expected to occur during partial degassing of HCl from aqueous solutions or a magmas. With this in mind, a survey of chlorine isotopes in volatile volcanic products was made. All samples are from Indonesian volcanoes. Indonesia lies at the triple junction of the Pacific plate, the Eurasian plate and the Australian plate (KATILI 1989, DE SMETH 1989). This complex geotectonic setting is also the reason for large seismic (RITSEMA et al. 1989, McCAFFREY 1989) and volcanic (KATILI 1975) activities.

The region was studied during the Indonesian-Dutch second Snellius Expedition (VAN HINTE & HARTONO 1989), which was a sequel to the Snellius Expedition in the years 1929-1930. This research has continued until present. One of the aims was the study of volcanism in relation with tectonic activity. Both rock (VAREKAMP et al. 1989, VAN BERGEN et al. 1989), and hot spring and fumarolic gas (POORTER et al. 1989a) samples were studied. Complex variations in subduction volcanism were found, which are partially

---

[1]Department of Geochemistry, Utrecht University, P.O.Box 80.021, 3508 TA Utrecht, The Neterlands

related to the (locally) subduction of Australian continental material under the oceanic crust (e.g. VROON 1992). Using He isotopes, it is possible to discriminate between subduction of continental (low values) and oceanic (high values) material (POREDA & CRAIG 1989). In these island volcanos $^3He/^4He$ ratios are low at the eastern volcanoes (Egon on Flores and Sirung on Pantar), and high at the western volcanoes, indicating a sharp transition between subducting oceanic material in the west and continental material in the east (HILTON & CRAIG 1989, SILVER 1989, HILTON et al. 1992). This trend (but more gradual) can also be distinguished with the aid of $^{87}Sr/^{86}Sr$, $^{18}O/^{16}O$ and $^{143}Nd/^{144}Nd$ isotope ratios (WHITFORD et al. 1977, WHITFORD & JAZEK 1979, MAGARITZ et al. 1978). The tectonic setting of these islands is described by HAMILTON (1979) and VAREKAMP et al. (1989).

In the present study, samples collected during Snellius in 1985 and two subsequent expeditions in 1989 and 1990 were measured for $\delta^{37}Cl$ variations. Samples include volcanic gases sampled in fumaroles (both as condensates and trapped in Giggenbach bottles), hot spring waters and water from crater lakes.

SAMPLE SITES

Samples were taken from volcanoes on six islands along the Sunda Arc (Java, Bali, Flores, Lomblen, Pantar and Alor; see FIG. 1). Most islands consist almost completely of volcanic rocks of arc-tholeiitic to potassic-alkaline composition (VAN BERGEN et al. 1989, VAREKAMP et al. 1989, WHELLER et al. 1987, WHITFORD 1975).

Fig. 1: Location of sampled volcanoes in Indonesia.

Condensates of volcanic gas were sampled at Ijen volcano (Java) and at Lewotolo volcano (Lomblen). Hot spring and crater lake samples are from Sirung (Pantar), Ijen (Java), Kelimutu and Egon (Flores), Batur (Bali), and the island of Alor (see **table 1**). Volcanic gasses trapped in Giggenbach bottles (GIGGENBACH & GOGUEL 1989) are from Merapi on Java (1 bottle, № G-18) and Lewotolo on Lomblen (the other 7 bottles).

The boiling spring and Sirung lake on Pantar are described by POORTER et al. (1989b). The Keli Mutu crater lake on Flores has been described by VAREKAMP & KREULEN (1990),

94

Table 1*: Isotope date of Indonesian water samples.*

Sample code	Description	Island	Volcano	$\delta^{37}Cl$	$\delta^{18}O$	δD
J1	Hot spring	Java	Ijen	0.19±0.07	-6.09	-49
J2	Crude lake	Java	Ijen	0.43±0.05	6.64	-14
J3	Condensate of gas	Java	Ijen	1.13±0.10	>10.00	>15
B1	Hot spring	Bali	Batur	-0.16±0.08	-5.71	-37
B2	Lake	Bali	Batur	-0.28±0.06	-1.41	-10
F1	KM-1	Flores	Kelimutu	-0.20±0.03	-0.86	-14
F2	KM-2	Flores	Kelimutu	-0.15±0.07	7.11	-2
F3	KM-3	Flores	Kelimutu	-0.23±0.10	-3.27	-23
F4	EGL-1	Flores	Egon	0.40±0.17	-2.00	-15
F5	EGL-3 (spring)	Flores	Egon	0.23±0.18	3.56	-7
L1	Condensate of gas	Lomblen	Lewotolo	0.25±0.07	6.05	-16
P1	Hot spring	Pantar	Sirung	0.07±0.09	5.50	-12
P2	Small lake	Pantar	Sirung	0.07±0.09	5.90	-13
P3	Crater lake	Pantar	Sirung	0.31±0.05	4.70	-3
P4	Crater lake	Pantar	Sirung	0.26±0.03	7.70	15
P5	Masi Aila	Pantar	Masi Aila	0.26±0.24	-3.90	-27
P6	Beach	Pantar	Beanggabang	-0.04±0.04	-3.50	-24
A1	Alor	Alor	-	-0.02±0.08	-3.60	-27

information on some of the other volcanos is given by POORTER *et al.* (1989a, 1991).

METHODS

Chloride in the samples is precipitated with $AgNO_3$ as $AgCl$. This $AgCl$ is reacted under vacuum with CH_3I to form CH_3Cl (TAYLOR & GRIMSRUD 1969, KAUFMANN 1984, see chapter 2). This gas is purified by gas chromatography and $^{37}Cl/^{35}Cl$ ratios are measured with a VG SIRA 24 Mass Spectrometer. $\delta^{18}O$ and δD are determined using conventional methods based on the isotopic equilibration with CO_2 and the reaction with hot uranium respectively.

Sample preparation for $\delta^{37}Cl$ analysis of the chloride collected in Giggenbach bottles is as follows: "Giggenbach bottles" contain approximately 50 ml of a 4N NaOH solution, in which the volcanic gases and their dissolved constituents can be trapped. An amount containing at least 0.1 mmol chloride is oxidized with 30% H_2O_2 at a temperature of about 80 °C to remove sulphides. Then the solution is diluted 100 times. Per 100 ml 6.00

gram KNO$_3$, 2.06 gram citric acid and 0.07 gram Na$_2$HPO$_4$.2H$_2$O is added. From this solution AgCl is precipitated. The precipitate is further treated as described above.

RESULTS

δ^{37}Cl, δ^{18}O and δD of the condensates and the water samples are shown in **table 1**. δ^{37}Cl varies from -0.23 to +1.13‰, δ^{18}O from -6.09 to +10‰ and δD from -49 to +15‰. Data on the volcanic gasses sampled in Giggenbach bottles are shown in **table 2**. δ^{37}Cl of

Table 2: Chloride concentration and δ^{37}Cl in Giggenbach bottles.

Sample	conc. Cl$^-$ (mg/ml)	δ^{37}Cl (‰ vs. SMOC)
G-11	2.5	0.03±0.09
G-18	0.3	3.24±0.20
G-22	0.3	9.47±0.37
G-24	3.1	-0.40±0.06
G-26	5.3	-1.55±0.06
G-28	7.1	0.57±0.18
G-31	2.2	0.50±0.05
G-35	1.6	1.66±0.04

these samples covers a very large range from -1.56 to +9.5‰, and shows an almost perfect inverse correlation with the chloride concentration in the Giggenbach bottles.

DISCUSSION

-Gas condensate and water samples

Based on their δ^{18}O and δD values (FIG. 2) the samples can be divided into meteoric waters, lying close to the meteoric water line, and geothermal waters, the latter being enriched in ^{18}O. The samples lying close to the meteoric water line represent water that was not modified by volcanic processes. Samples with substantially increased δ^{18}O values (further referred to as geothermal waters) have exchanged oxygen isotopes with the surrounding rocks at high temperatures (e.g. CRAIG 1963, see HOEFS 1987) and/or have been modified by evaporation at elevated temperatures (e.g. the crater lakes).

The chlorine isotopes are plotted against δ^{18}O in FIG. 3. There is a striking separation between the meteoric waters with slightly negative δ^{37}Cl values and the geothermal waters with slightly positive δ^{37}Cl values. These differences may be related to the origin of the chloride in the samples. Meteoric waters will contain chloride from weathered volcanic rocks and possibly also seawater spray. Since most meteoric water samples have

96

Fig. 2: $\delta^{18}O$ vs. δD plot of volcanic gases, springs and crater lakes. Triangles are samples with negative $\delta^{37}Cl$ and $\delta^{18}O$ values, dots are samples with positive $\delta^{37}Cl$ and $\delta^{18}O$ values, diamonds are samples with negative $\delta^{37}Cl$ and positive $\delta^{18}O$ values and squares are samples with positive $\delta^{37}Cl$ and negative $\delta^{18}O$ values.

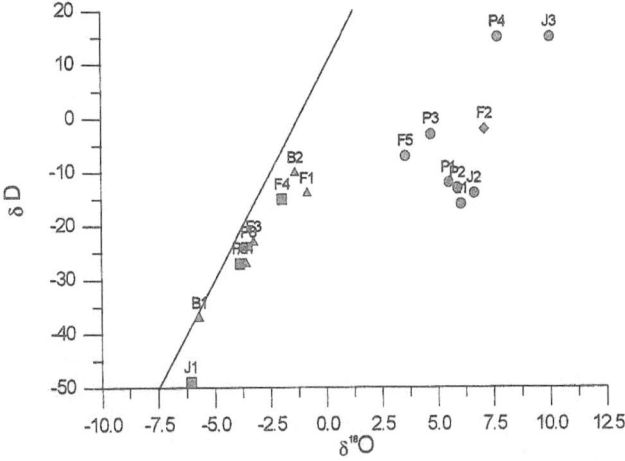

Fig. 3: $\delta^{37}Cl$ vs. $\delta^{18}O$ plot of volcanic gases, springs and crater lakes. Symbols have the same meanings as in Fig. 2.

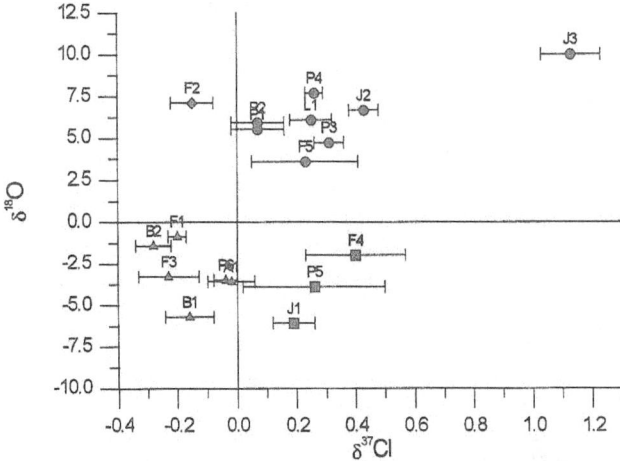

negative $\delta^{37}Cl$ values it is suggested that the volcanic rocks have negative $\delta^{37}Cl$ values. The geothermal samples, however, probably derived a substantial part of their chloride from HCl in volcanic gas. This would imply that HCl in volcanic gases has a positive $\delta^{37}Cl$ whereas the outgassed lavas have negative values.

This might be in agreement with experiments by HOERING & PARKER (1961). They measured the chlorine isotope exchange according to the reaction:

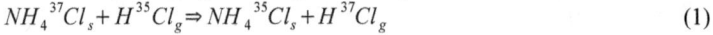

$$NH_4^{37}Cl_s + H^{35}Cl_g \Rightarrow NH_4^{35}Cl_s + H^{37}Cl_g \qquad (1)$$

and found that the fractionation factor α at 200 °C was:

$$\alpha = \frac{\left(\dfrac{^{37}Cl}{^{35}Cl}\right)_B}{\left(\dfrac{^{37}Cl}{^{35}Cl}\right)_A} = 1.0009 \pm 0.0003 \qquad (2)$$

Thus HCl in equilibrium with NH₄Cl will be enriched in ^{37}Cl. If a similar relation should exist in silicate melt-HCl equilibria, then the HCl escaping in volcanic gases will indeed be enriched in ^{37}Cl.

Experiments by HOWALD (1960) gave very confusing results. HOWALD (1960) measured the chlorine isotope effect of HCl gas in equilibrium with solutions of inorganic chlorides in glacial acetic acid (= 99-100% acetic acid). For pure HCl he found that the fractionation was dependent on the water concentration in the acetic acid. At water concentrations below 1.5M, gaseous HCl was depleted in ^{37}Cl, at higher water concentrations gaseous HCl was enriched in ^{37}Cl. In experiments with LiCl and SrCl₂ the fractionation factor depended on the fraction HCl dissolved in the liquid phase. For HCl fractions lower than about 0.5 the HCl gas was enriched in ^{37}Cl, for fractions higher than 0.5 the solution was enriched in ^{37}Cl. For the time being it is not clear how these results can be extrapolated to volcanic gas-rock systems, but at least they show that escaping HCl may be fractionated.

In chapter 12 an experiment with HCl escaping according to the reaction:

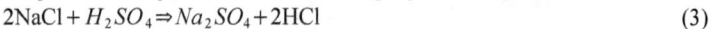

$$2NaCl + H_2SO_4 \Rightarrow Na_2SO_4 + 2HCl \qquad (3)$$

is described. The $\delta^{37}Cl$ of the residue (NaCl containing Na₂SO₄) was 0.52‰ higher than $\delta^{37}Cl$ in the original NaCl. Thus in this case HCl with a lower $\delta^{37}Cl$ must have escaped.

Clearly, more research on this subject must be done to understand these strange and sometimes contradictory results. The $\delta^{37}Cl$ measurements reported in this chapter agree with the experiments by HOERING & PARKER (1961) in which gaseous HCl is enriched in $\delta^{37}Cl$ relative to the residue. This also seems to agree with results on primary carbonatites (chapter 13), that were found to have slightly negative $\delta^{37}Cl$ values.

Four samples have $\delta^{37}Cl$ values that do not agree with their $\delta^{18}O$ and δD values. Sample F2 (KM-2) has a negative $\delta^{37}Cl$ while $\delta^{18}O$ and δD point to a geothermal origin. This sample, however is from a crater lake and it is very likely that the high $\delta^{18}O$ and δD are caused by evaporation. Chlorine is then mainly derived from the rocks and thus has a low $\delta^{37}Cl$ value. Samples F4 (EGL-1), P5 (Masi Aila) and J1 (Ijen Hot Spring) have high $\delta^{37}Cl$ values while their $\delta^{18}O$ and δD signals points to meteoric water. In these cases it is supposed that original geothermal waters were diluted with fresh meteoric water so that

Fig. 4: Relationship between Cl⁻ and $\delta^{37}Cl$ in samples collected in Giggenbach bottles.

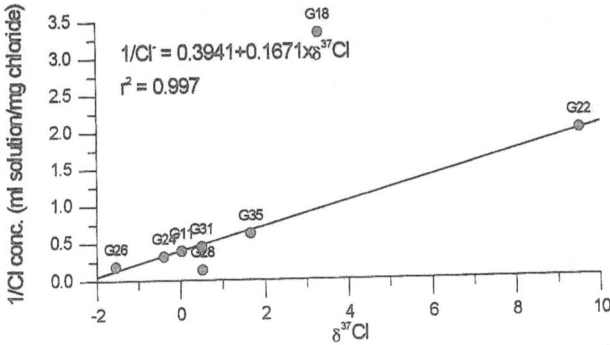

Fig 5: Test for $\delta^{37}Cl$ measurements in very concentrated solutions. Dots are measured values for samples collected in Giggenbach bottles, open circles are concentrated test samples.

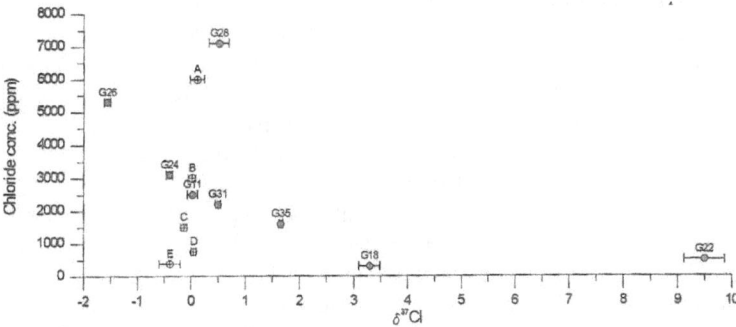

the H_2O got a meteoric water signal whereas most of the chloride is still of volcanic gas origin.

-Giggenbach bottles

In the Giggenbach bottles $\delta^{37}Cl$ values cover an extreme range from -1.56 to +9.5‰. The value of +9.5‰ is the highest ever measured. A very good correlation exists between $\delta^{37}Cl$ and the inverse of the chloride concentration (FIG. 4). Two samples do not fit, G-28 with a too high chloride concentration and G-18 with a too low concentration relative to the $\delta^{37}Cl$ regression line. The problem with G-28 may be related to the fact that this Giggenbach bottle was broken and the sample was stored in a glass bottle. In G-18 no total gas yield was determined. This also is the only bottle from Merapi (Java) and this may be a cause for the deviating $\delta^{37}Cl$ value. Thus only two samples from a different location, or that were broken do not plot on the correlation line. Because of the very large range in $\delta^{37}Cl$ values, it seems unlikely that these results have a geological significance. The variations may be due to analytical problems related to the chemistry of these Giggenbach bottles, or to incomplete sampling of the volcanic gas. To test this, 4N NaOH solutions containing 750 to 12000 ppm chloride were prepared and acidified with 65% HNO_3 so that the concentrations were halved to 375 to 6000 ppm. In five test runs (A, B, C, D, E, FIG. 5) $\delta^{37}Cl$ was about 0‰ (except for the sample with only 375 ppm chloride, which was too small). Although other analytical problems may have caused the variations, it is also possible that fractionations occur during sampling. For this reason it is recommended to reevaluate both the $\delta^{37}Cl$ determinations and the chemical analyses on Giggenbach bottles.

CONCLUSIONS

Two main groups can be recognized in volcanic water samples. Those which contain meteoric water have negative $\delta^{37}Cl$ values, and those containing geothermal water have positive $\delta^{37}Cl$ values. Chloride in the meteoric waters is derived from weathered rock and seawater spray, chloride in the geothermal water derived for a substantial part from HCl in volcanic gases. Although experimental results are unclear, it seems that HCl in equilibrium with dissolved chloride (in this system) is enriched in ^{37}Cl. The rocks from which the HCl escaped thus will remain with lower $\delta^{37}Cl$ values. This seems to be comparable with measurements in carbonatites (chapter 13), as it is concluded that primary mantle carbonatites have negative $\delta^{37}Cl$ values, which might indicate that HCl escaped from the mantle was enriched in ^{37}Cl.

In contrast with the water samples, $\delta^{37}Cl$ values in Giggenbach bottles probably have no geological meaning. $\delta^{37}Cl$ variations are dependent on the chloride concentration in the bottle. Because of this striking result it is recommended that the chemical and other analyses of Giggenbach bottles are critically reviewed and test if they are not the result of sampling effects, or analytical errors.

ACKNOWLEDGEMENTS

Samples are provided by René Poorter, Johan Varekamp and Jurian Hoogewerff. Jan Meesterburrie, Anita van Leeuwen-Tolboom and Arnold van Dijk measured $\delta^{18}O$ and δD of these samples. S.O. Scholten carefully read an earlier version of this chapter and suggested many improvements.

REFERENCES

CRAIG H. (1963) The isotopic chemistry of water and carbon in geothermal areas. In: *Nuclear geology of geothermal areas.* 17-53. Spoleto, Sept 9-13, 1963

DE SMETH M.E.M. (1989) A geometrically consistent plate-tectonic model for Eastern Indonesia. *Neth. J. Sea Res.* **24** 173-183

EASTOE C.J. & GUILBERT J.M. (1992) Stable chlorine isotopes in hydrothermal processes. *Geochim. Cosmochim. Acta* **56** 4247-4255

GIGGENBACH W.F. & GOGUEL R.L. (1988) Methods for the collection and analysis of geothermal and volcanic gas samples. *Chem. Div. S. I. R. Petone NZ. Rept.CD 2387.*

HAMILTON W. (1979) Tectonics of the Indonesian region. *USGS Prof. Pap.* **1078** 345 pp.

HILTON D.R. & CRAIG H. (1989) A helium isotope transect along the Indonesian archipelago. *Nature* **342** 906-908

HILTON D.R., HOOGEWERFF J.A., VAN BERGEN M.J. & HAMMERSCHMIDT K. (1992) Mapping magma sources in the east Sunda-Banda arcs, Indonesia: Constraints from helium isotopes. *Geochim. Cosmochim. Acta* **56** 851-859

HOEFS J. (1987) Stable isotope geochemistry. Third, completely revised and enlarged edition. *Minerals and Rocks* **9** 241 pp.

HOERING T.C. & PARKER P.L. (1961) The geochemistry of the stable isotopes of chlorine. *Geochim. Cosmochim. Acta* **23** 186-199

HOWALD R.A. (1960) Ion Pairs. I. Isotope effects shown by chloride solutions in glacial acetic acid. *J. Amer. Chem. Soc.* **82** 20-24

KATILI J.A. (1975) Volcanism and plate tectonics in the Indonesian island arcs. *Tectonophysics* **26** 165-188

KATILI J.A. (1989) Review of past and present geotectonic concepts of eastern Indonesia. *Neth. J. Sea Res.* **24** 103-129

KAUFMANN R.S. (1984) *Chlorine in groundwater: Stable isotope distribution.* Ph.D. Thesis, University of Arizona. 137 pp.

MAGARITZ M., WHITFORD D.J. & JAMES D.E. (1978) Oxygen isotopes and the origin of high $^{87}Sr/^{86}Sr$ andesites. *Earth Planet. Sci. Lett.* **40** 220-230

MCCAFFREY R. (1989) Seismological constraints and speculations on Banda arc tectonics. *Neth. J. Sea Res.* **24** 141-152

POORTER R.P.E., VAREKAMP J.C., SRIWANA T., VAN BERGEN M.J., ERFAN R.D., SUHARYONO K., WIRAKUSUMAH A.D. & VROON P.Z. (1989a) Geochemistry of hot springs and fumarolic gases from the Banda arc. *Neth. J. Sea Res.* **24** 323-331

POORTER R.P.E., VAREKAMP J.C., VAN BERGEN M.J., KREULEN R., SRIWANA T., VROON P.Z. & WIRAKUSUMAH A.D. (1989b) The Sirung volcanic boiling spring: An extreme chloride-rich, acid brine on Pantar (Lesser Sunda Islands, Indonesia). *Chem. Geol.* **76** 215-228

POORTER R.P.E., VAREKAMP J.C., POREDA R.J., VAN BERGEN M.J. & KREULEN R. (1991) Chemical and isotopic compositions of volcanic gases from the east Sunda and Banda arcs, Indonesia. *Geochim. Cosmochim. Acta* **55** 3795-3807

POREDA R. & CRAIG H. (1989) Helium isotope ratios in circum-Pacific volcanic arcs. *Nature* **338** 473-478

RITSEMA A.R., SUDARMO R.P. & PUTU PUTJA I. (1989) The generation of the Banda arc on the basis of its seismicity. *Neth. J. Sea Res.* **24** 165-172

SILVER E.A. (1989) Helium line in the Banda arc. *Nature* **342** 856-857

TAYLOR J.W. & GRIMSRUD E.P. (1969) Chlorine isotope ratios by negative ion mass spectrometry. *Anal. Chem.* **41** 805-810

VAN HINTE J.E. & HARTONO H.M.S. (1989) Executive summary of theme I Snellius-II Expedition. 'Geology and geophysics of the Banda arc and adjacent areas'. *Neth. J. Sea Res.* **24** 95-102

VAN BERGEN M.J., ERFAN R.D., SRIWANA T., SUHARYONO K., POORTER R.P.E., VAREKAMP J.C., VROON P.Z. & WIRAKUSUMAH A.D. (1989) Spatial geochemical variations of arc volcanism around the Banda Sea. *Neth. J. Sea Res.* **24** 313-322

VAREKAMP J.C. & KREULEN R. (1990) Exotic fluids: Chemistry of the Keli Mutu crater lakes, Flores, Indonesia. *Eos* **71** 1674

VAREKAMP J.C., VAN BERGEN M.J., VROON P.Z., POORTER R.P.E., WIRAKUSUMAH A.D., ERFAN R., SUHARYONO K. & SRIWANA T. (1989) Volcanism and tectonics in the Eastern Sunda arc, Indonesia. *Neth. J. Sea Res.* **24** 303-312

VROON P.Z. (1992) Subduction of continental material in the Banda Arc, Eastern Indonesia. Sr-Nd-Pb isotope and trace-element evidence from volcanics and sediments. *Geol. Ultraj.* **90** 205 pp. Ph.D. thesis, Utrecht University.

WHELLER G.E., VARNE R., FODEN J.D. & ABBOT M.J. (1987) Geochemistry of quaternary volcanism in the Sunda-Banda Arc, Indonesia, and three-component genesis of island arc basaltic magmas. *J. Volcan. Geotherm. Res.* **32** 137-160

WHITFORD D.J. (1975) Strontium isotopic studies of the volcanic rocks of the Sunda Arc, Indonesia, and their petrogenetic implications. *Geochim. Cosmochim. Acta* **39** 1287-1302

WHITFORD D.J. & JAZEK P.A. (1979) Origin of late-Cenozoic lavas from the Banda arc, Indonesia: Trace elements and Sr isotope evidence. *Contrib. Mineral. Petrol.* **68** 141-150

WHITFORD D.J., COMPSTON W., NICHOLLS I.A. & ABBOTT M.J. (1977) Geochemistry of Late Cenezoic lavas from eastern Indonesia: Role of subducted sediments in petrogenesis. *Geology* **5** 571-575

CHAPTER 9

Other Measurements of Chlorine Stable Isotopes in Aqueous Systems

H.G.M. Eggenkamp[1]

ABSTRACT-- In order to make an inventory of chlorine isotope differences in different environments chlorine isotope ratios are measured in five small sample sets of natural waters. In these samples only few as yet explained variations were found. Further research on these samples could be very interesting.

MINERAL WATERS RELATED TO THE RIBAMA FAULT IN NORTHERN PORTUGAL

-Introduction

On the Portuguese mainland many mineral springs are found. These waters have been used for medical purposes since pre-Roman times (DE MENEZES CORREA ACCIAIUOLI 1952). All springs are related to fault systems (SAN-BENTO MENEZES & RODRIGUES DA SILVA 1988). Although many studies were carried out (e.g. CABRAL *et al.* 1977, HERCULANO DE CARVALHO 1966, AIRES-BARROS 1979, MENEZES DE ALMEIDA 1982), isotope data for the Portuguese spring waters are scarce. In this section chlorine isotope measurements are presented for seven springs in the northernmost part of Portugal. All are connected to the Ribama fault, a major fault in Northern Portugal with a SSW-NNE direction between Coimbra and Chaves (RIBEIRO *et al.* 1972). The samples are divided in three groups, two north of Chaves, three in the surroundings of Pedras Salgadas and two near Saô Pedro do Sul.

-Samples

The samples are collected in the summer of 1985 during a fieldwork on the hydrogeochemistry of the northern area (EGGENKAMP *et al.* (1987). Carvalhal and Saô Pedro do Sul are sulphide waters, the others are bicarbonate waters waters (DE MENEZES CORREA ACCIAUOLI 1952). The regions in which the springs are found lie about 130 km apart.

-Methods

Because of the low Cl⁻ content, according to EGGENKAMP *et al.* (1987) it is from 35 to 50 ppm, for each measurement 100 ml sample is used. To this solution dry chemicals are added as described in chapter 2.

---

[1]Department of Geochemistry, Utrecht University, P.O.Box 80.021, 3508 TA Utrecht, The Netherlands

Table 1: Cl⁻ and $\delta^{37}Cl$ of Portuguese springs.

Spring	Distance from the north (km)	Cl⁻ (ppm)	$\delta^{37}Cl$ (‰)
Vilarelho da Raia	0	30	-0.03±0.04
Chaves	11	53	0.03±0.14
Campilho	21	13	-0.07±0.11
Sabroso	31	34	0.10±0.08
Pedras Salgadas	35	48	0.22±0.12
Carvalhal	110	28	-0.08±0.04
São Pedro do Sul	123	28	0.00±0.03

-Results and discussion

As can be seen in **table 1**, almost all $\delta^{37}Cl$ values are close to SMOC (between -0.08 and +0.10‰). One sample is significantly different from SMOC (Pedras Salgadas, 0.22‰).

Depending on their location the samples can be divided in three groups. The northern group contains Chaves and Vilarelho da Raia, the middle group Campilho, Sabroso and Pedras Salgadas, and the southern group Carvalhal and Saõ Pedro do Sul. If $\delta^{37}Cl$ of the samples is plotted versus their position along the fault it is found that within each group the southernmost sample has the most positive $\delta^{37}Cl$ and the northernmost sample has the most negative $\delta^{37}Cl$ (Fig. 1). Since diffusion is a likely mechanism for fractionating chlorine isotopes it is suggested that diffusion occurs preferentially northwards. For each group of samples the chloride concentration decreases from south to north and $\delta^{37}Cl$ decreases at the same time. An other possible explanation for the variations can be hydrothermal alteration. Since the number of samples is only small, the observed relation may be coincidence. More research is recommended.

REFERENCES

AIRES-BARROS L. (1979) Termometria geoquímica. Princípios gerains, aplicações geotérmicas a casos portugueses. *Communic. Serv. Geol. Port.* **64** 103-132

CABRAL J.M.P., HERCULANO DE CARVALHO A. & LIMA M.B. (1977) Aplicação de métodos de taxonomia numérica na classificação de águas minerais de Portugal continental. *Communic. Serv. Geol. Port.* **61** 343-363

DE MENEZES CORREA ACCIAUOLI L. (1952) *Le Portugal hydromineral.* I volume. Direction Général des Mines et des Services Géologiques. Lisbonne. 284 pp.

EGGENKAMP H.G.M., WIJLAND G.C. & SAAGER P. (1987) *Hydrogeochemical study of the Chaves-Vila-Pouca area.* Unpubl. MSc. Study.

HERCULANO DE CARVALHO A. (1966) Caracterização físico-química expedita das águas minerais

Fig. 1: Results δ³⁷Cl measurements of Portuguese springs, indicating the three different groups

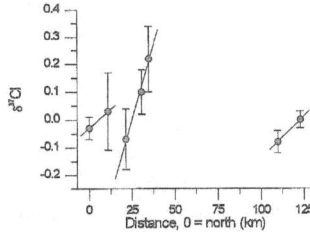

portuguesas. *Bolm. Acad. Ciênc. Lisb.* **38** 30-40

MENEZES DE ALMEIDA F. (1982) Novos dados geothermométricos sobre águas de Chaves e de São Pedro do Sul. *Communic. Serv. Geol. Port.* **68** 179-190

RIBEIRO A, CONDE L. & MONTEIRO E.J. (1972) *Carta tectonica de Portugal. Escala 1:1,000,000.* Instituto Geografico e Cadastral. Lisboa.

SAN-BENTO MENEZES M. & RODRIGUES DA SILVA A.M. (1988) Development of geothermal resources in Portugal. *Geothermics* **17** 565-574*Ch. 9: Other measurements in aqueous systems*

Fig. 2: Location IJsselmeer surface samples.

IJSSELMEER SURFACE WATER (THE NETHERLANDS)

-Introduction

The IJsselmeer is a 1700 km² large artificial lake in the centre of the Netherlands (see chapter 4). It is filled with fresh water from the river IJssel and drainage water from the surrounding polders and higher sand grounds. The lake is divided in two parts by a dyke between the cities of Enkhuizen and Lelystad.

Eight surface water samples were taken on a single day from several parts of the lake (Fig. 2). Also, two samples were taken from the adjacent Wadden Sea and one on the River IJssel near Deventer.

-Results and discussion

Results are given in **table 2** and Fig. 3. $\delta^{37}Cl$ varies from slightly negative values (-0.11 and -0.07‰) in the saline Wadden Sea samples, to slightly positive values in the fresh lake and river samples (0.00 to 0.16‰). The difference in $\delta^{37}Cl$ between the IJsselmeer and the Wadden Sea samples is barely significant.

Table 2: *Results $\delta^{37}Cl$ measurements surface water IJsselmeer. Sample numbers refer to the locations of Fig. 2*

Sample	$\delta^{37}Cl$ (‰)
2650	0.08±0.05
2651	0.14±0.00
2652	0.05±0.04
2653	0.12±0.05
2654	0.10±0.02
2655	0.14±0.16
2656	0.05±0.00
2657	0.16±0.02
2658	0.00±0.06
2659	-0.11±0.14
2660	-0.07±0.03

Fig. 3: *Results $\delta^{37}Cl$ measurements IJsselmeer samples.*

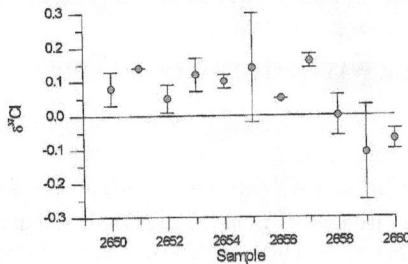

DUTCH TAP- AND GROUND WATER

-Introduction

Four dutch tapwater and one groundwater samples were measured. The four tapwaters are from Utrecht, Amersfoort, Hilversum and Laren. The groundwater sample is from Laren.

-Results and discussion

As can be seen in **table 3** the tapwaters of Amersfoort, Hilversum and Laren do not deviate significantly from S.M.O.C., and this is also the case for the groundwater sample from Laren. The tapwater from the university centre at Utrecht (de Uithof), however, deviates significantly; its $\delta^{37}Cl$ is -0.49. The groundwater used to produce this water may have a negative $\delta^{37}Cl$, in which case it is concluded that $\delta^{37}Cl$ variations occur in the dutch subsoil. Such variations can be produced by diffusion during groundwater movement and may therefore be of interest for hydrologic research. An alternative explanation is that the deviating $\delta^{37}Cl$ of the Utrecht tapwater is an artfact produced by processing in the water plant.

Table 3: Results $\delta^{37}Cl$ measurements dutch tapwater.

Sample	$\delta^{37}Cl$ (‰)
Groundwater Laren	0.02±0.02
Tapwater Laren	0.00±0.01
Tapwater Hilversum	-0.05±0.03
Tapwater Amersfoort	0.15±0.19
Tapwater Utrecht	-0.49±0.19

LITHIUM BRINES

-Introduction

Samples from three lithium rich hypersaline lakes, Salar de Uyuni in Bolivia, Salar de Atacama in Chile and Silver Peak-Clayton Valley in Nevada, U.S.A. (KUNASZ 1980) were measured for chlorine isotopes. All three samples are saturated with chloride.

The brines occur in dry regions and are associated with volcanic activity which is probably the source of the lithium. Chlorine isotopes sometimes deviated from SMOC in volcanic systems (see chapter 7). Isotopic inhomogenities may further be enhanced by the processes that concentrate the lithium. For these reasons the lithium brines are promising to look for $\delta^{37}Cl$ variations. In the brines of Salar de Uyuni large variations in lithium content occur (from 90 to 1150 ppm, ERICKSEN *et al.* 1978), showing that the water is not well mixed, of course also because precipitation of salt occurs within the lake.

Table 4: *Results $\delta^{37}Cl$ measurements lithium brines.*

Salar	Country	$\delta^{37}Cl$ (‰)
Salar de Atacam	Chili	-0.18±0.07
Salar de Uyuni	Bolivia	-0.45±0.13
Silver Peak	Nevada, U.S.A.	0.36±0.22

-Results and discussion

The three samples vary from -0.45 to +0.36‰ (see **table 4**) suggesting that these special environments merit a more detailed study.

-References

ERICKSEN G.E., VINE J.D. & RAUL BALLÓN A. (1978) Chemical composition and distribution of lithium-rich brines in Salar de Uyuni and nearby salars in southwestern Bolivia. *Energy* **3** 355-363

KUNASZ I.A. (1980) Lithium in brines. *Proc. Fifth Int. Symp. Salt* **1** 115-117

DEEP-SEA BRINES

-Introduction

Three samples are measured from two Mediterranean anoxic hypersaline basins. These basins are the Tyro basin, discovered in 1983 (DE LANGE & TEN HAVEN 1983) and the Bannock basin, discovered in 1984 (SCIENTIFIC STAFF OF CRUISE BANNOCK 1984-12 1985, FIG. 4). In these basins salinity of the water increases sharply at a depth of about 3350 meter. The reason for the high salinity is that Messinian salt deposits (Miocene) that underlie most of the Mediterranean (NESTEROFF 1973, HSÜ *et al.* 1973) are dissolved in the seawater. A description of the basins is given in CAMERLENGHI (1990). The chemical composition is described by DE LANGE *et al.* (1990). Almost no mixing occurs between the two water masses. Since diffusion then is a possible way of moving it must be possible to detect this with chlorine isotopes.

Fig. 4: Locations Tyro and Bannock basins in the Mediterranean sea (Henneke 1993).

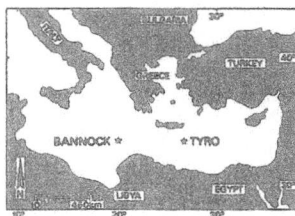

-Samples

The three samples are ABC-8/10 from the Tyro basin from a depth of about 800 meter. Its chloride concentration is about 0.61 M (or 22000 ppm). From the Bannock basin are measured ABC-38/17 and ABC-53/6, from depths of about 1372 and 3730 meter and chlorinities of 0.60 M (21000 ppm) and 5.07 M (180000 ppm) according to Ullman *et al.* (1990).

Table 5: Results $\delta^{37}Cl$ measurements deep-sea-brines.

Sample	$\delta^{37}Cl$ (‰)
ABC-8, Tyro-basin	0.18±0.16
ABC-38, Bannock-basin	0.32±0.02
ABC-53, Bannock-basin	-0.05±0.15

-Results and discussion

Results (**table 5**) are difficult to interpret. The value next to zero is the only hypersaline sample and the two measured normal seawaters are relatively positive. It is supposed that more samples will be measured for $\delta^{37}Cl$. Especially around the saline-hypersaline interface this can be very interesting because here possibly can be determined on what scale diffusion appears. From measurements in the hypersaline layer possible differences between the Tyro and Bannock basins can be determined.

-References

Camerlenghi A. (1990) Anoxic basins of the eastern Mediterranean: geological framework. *Mar. Chem.* **31** 1-19

DE LANGE G.J. & TEN HAVEN H.L. (1983) Recent sapropel formation in the eastern Mediterranean. *Nature* **305** 797-798

DE LANGE G.J., MIDDELBURG J.J., VAN DER WEIJDEN C.H., CATALANO G., LUTHER III G.W., HYDES D.J., WOITTIEZ J.W.R. & KLINKHAMMER G.P. (1990) Composition of anoxic brines in the Tyro and Bannock Basins, eastern Mediterranean. *Mar. Chem.* **31** 63-88

HENNEKE E. (1993) Early diagenetic processes and sulphur speciation in pore waters and sediments of the hypersaline Tyro and Bannock Basins, eastern Mediterranean. *Geol. Ultrai.* **108** 150 pp. Ph.D. thesis, Utrecht University.

HSÜ K.J., CITA M.B. & RYAN W.B.F. (1973) The origin of the Mediterranean evaporites. **In** RYAN W.B.F., HSÜ K.J. *et al.* (Editors), *Initial reports DSDP* **13** U.S. Government Printing Office pp 673-694

NESTEROFF W.D. (1973) Mineralogy, petrography, distribution, and origin of the Mediterranean evaporites. **In** RYAN W.B.F., HSÜ K.J. *et al.* (Editors), *Initial reports DSDP* **13** U.S. Government Printing Office pp. 928-937

SCIENTIFIC STAFF OF CRUISE BANNOCK 1984-12 (1985) Gypsum precipitation from cold brines in an anoxic basin in the eastern Mediterranean. *Nature* **314** 152-154

ULLMAN W.J., LUTHER III G.W., DE LANGE G.J. & WOITTIEZ J.R.W. (1990) Iodine chemistry in deep anoxic basins and overlying waters of the Mediterranean sea. *Mar. Geol.* **31** 153-170

CHAPTER 10

Chlorine Stable Isotope Fractionation in Evaporites

H.G.M. Eggenkamp[1], R. Kreulen[1] and A.F. Koster van Groos[2]

ABSTRACT-- Chlorine isotope fractionation ($^{37}Cl/^{35}Cl$) between NaCl, KCl and $MgCl_2.6H_2O$ and their saturated solutions was determined in laboratory experiments at 20 °C. The results are as follows:

NaCl - solution	+0.235±0.073‰
KCl - solution	-0.050±0.087‰
$MgCl_2.6H_2O$ - solution	-0.067±0.096‰

These data were used to approximate the isotope fractionation factors of kainite ($K_4Mg_4Cl_4(SO_4)_4.11H_2O$) and carnallite ($KMgCl_3.6H_2O$). From these the stable chlorine isotope fractionation during the formation of evaporite was calculated, using a Rayleigh fractionation model. The model predicts that $\delta^{37}Cl$ of the precipitate decreases systematically during the main phase of halite crystallization, but increases again at the latest stage of evaporation. The chlorine isotope fractionation model was tested on a core from the upper Zechstein III salt formation, which includes multiple evaporation cycles. The salt core contains layers dominated by halite and by potassium-magnesium salts. The potassium-magnesium salts represent the final stages of evaporation, and contain up to 75% carnallite ($KMgCl_3.6H_2O$) and bischofite ($MgCl_2.6H_2O$). The observed chlorine isotope fractionation is in general agreement with the Rayleigh fractionation model. During the main crystallization phase of halite, $\delta^{37}Cl$ decreases substantially, but this trend reverses when Mg-salts such as bischofite begin to crystallize. It is concluded that $\delta^{37}Cl$ can be used as a monitor of evaporation cycles. Thus, it provides quantitative information on the proportion of salt that has been deposited, on the input of fresh seawater and on the disturbance by post-depositional processes.

INTRODUCTION

Chlorine isotopes have been largely ignored in evaporite studies, such in contrast to the isotopes of sulphur and oxygen in evaporitic sulphates. The latter show significant isotope fractionation, as they are affected by reduction of sulphate by bacteria, and other biological effects (AULT & KULP 1959, ERIKSSON 1963, NIELSEN 1966, REES 1970). During evaporation, some non-biological fractionation of the sulphur isotopes occurs also, and the isotope ratio of the precipitates is slightly different when compared to the seawater from which the evaporite was formed. This effect is illustrated by the consistently lower $\delta^{34}S$ values found in K-Mg sulphate facies relative to the gypsum and anhydrite facies (NIELSEN & RICKE 1964, HOLSER & KAPLAN 1966). THODE & MONSTER (1965) determined from both new experimental and old calculated data that the sulphur isotope fractionation between dissolved sulphate and the precipitated gypsum is 1.00165±0.00012. These $\delta^{34}S$ differences are quite small when compared to the differences caused by biological effects,

---

1Department of Geochemistry, Utrecht University, P.O.Box 80.021, 3508 TA Utrecht, The Netherlands
2Department of Geological Sciences, The University of Illinois at Chicago, Box 4348, Chicago, Illinois 60680, USA

and not much attention has been given to non-biological fractionation. In contrast, chlorine does not take part in biochemical reactions. Therefore, any chlorine isotope fractionation found in evaporites must have occurred during the crystallization of the various salts. The following study reports on a series of experiments on the fractionation of chlorine isotopes in brines and evaporites. The results show that the total fractionation of chlorine isotopes in evaporites, although small, is significant. Comparison of $\delta^{37}Cl$ data from these experiments and a core from the upper Zechstein III shows that the evaporation state and history of an evaporite sequence can be evaluated.

-Previous work on $\delta^{37}Cl$ in evaporites

Very little work has been done on chlorine isotopes in evaporites. Early chlorine isotope studies lacked the analytical precision to detect the small variations that were inferred for $\delta^{37}Cl$. HOERING & PARKER (1961) measured a selection of evaporite samples but found no variations outside their limits of precision. Accompanying experiments on the isotope exchange between NaCl crystals and a saturated solution, indicated a fractionation factor of 1.0002 ± 0.0003 from which it was concluded that no significant fractionation occurred. Since the early eighties, the precision mass spectrometers have become better and sample preparation procedures are greatly improved. A few evaporite samples were measured by KAUFMANN et al. (1984, 1988). They found a $\delta^{37}Cl$ of +0.19‰ in halite from the Weeks Island Dome, Louisiana and +0.52‰ in the bedded salts of the Salina Formation, Ohio.

VENGOSH et al. (1989) used negative thermal-ionization mass spectrometry to measure $\delta^{37}Cl$ variations. In evaporites from inland ponds in China and Australia they found rather extreme $\delta^{37}Cl$ values, ranging from -7.2 to +24.7‰, with a large standard deviation of 0.7 to 2‰. Their results have not been confirmed by conventional analytical methods. Also, they did not find systematic fractionation trends, although the more evaporated samples tend to deviate more, either in positive or in negative direction.

Recently we improved our analytical technique, which allows us to measure $\delta^{37}Cl$ with a precision that is sufficient to determine $\delta^{37}Cl$ of a salt and a coexisting solution during different stages of evaporation.

METHODS

Evaporation experiments were performed in which $\delta^{37}Cl$ of NaCl, KCl, and $MgCl_2.6H_2O$, and of the coexisting aqueous solutions were measured. The results were used to calculate theoretical fractionation trends of evaporating seawater. Next, the results were compared to a series of samples from the upper Zechstein III evaporite formations from the northern Netherlands.

-Analytical method

The chlorine isotope ratios are measured on chloromethane gas (TAYLOR & GRIMSRUD

112

1969, KAUFMANN 1984), using a slightly modified method. Salt samples were dissolved in water and the chloride is quantitatively precipitated by adding $AgNO_3$. The AgCl precipitate and an excess amount of CH_3I are sealed together in an evacuated glass tube and allowed to react for 48 hours at 75 °C to form CH_3Cl. Next, CH_3Cl and the remaining CH_3I are separated by gas chromatography and the purified CH_3Cl is measured on a VG SIRA-24 mass spectrometer equipped with adjustable collectors. The trap current is reduced to 100 µA in order to bring the minor beam down to values $<10^{-10}$A that can be handled by the mass spectrometer while still maintaining sufficient gas pressure in the inlet system.

$\delta^{37}Cl$ values are reported relative to the chlorine isotopic composition of seawater (SMOC = Standard Mean Ocean Chloride) which is a very constant and well mixed reservoir (KAUFMANN 1984).

-Evaporation experiments

Near-saturated solutions of reagent grade NaCl, KCl and $MgCl_2.6H_2O$ were allowed to evaporate at ambient temperature. A precipitate started to form after two days for KCl, after three days for NaCl and after 28 days for $MgCl_2.6H_2O$. The solutions and the precipitate were separated by decanting and filtering. The precipitate was rinsed with acetone to remove the remaining solution. The solutions and the salts were measured for chlorine isotopes as described above.

EXPERIMENTAL DETERMINATION OF CHLORIDE ISOTOPE FRACTIONATION

-Results of evaporation experiments

Two parallel evaporation experiments were performed for each of the three components NaCl, KCl and $MgCl_2.6H_2O$. In each experiment the precipitate and the solution were measured several times, and the data were combined to calculate equilibrium fractionation factors for these salts. The fractionation factor of NaCl with respect to the brine solution, +0.235±0.073‰, is strongly positive. For both KCl and $MgCl_2.6H_2O$, the fractionation factors, respectively -0.050±0.087‰ and -0.067±0.096‰ are slightly negative. The uncertainty is defined as the standard deviation of the $\delta^{37}Cl$ differences between crystal and solution of all measurements. The standard deviations are within normal values; more precise values for the fractionation requires a much more careful, very difficult evaporation experiment than was carried out in this study.

-Precipitation of salt from seawater

The sequence of salt minerals that precipitate upon evaporation of standard seawater is as follows: gypsum ($CaSO_4.2H_2O$), halite (NaCl), bloedite ($Na_2Mg(SO_4)_2.4H_2O$), epsomite ($MgSO_4.7H_2O$), kainite ($K_4Mg_4Cl_4(SO_4)_4.11H_2O$), hexahydrite ($MgSO_4.6H_2O$), kieserite ($MgSO_4.H_2O$), carnallite ($KMgCl_3.6H_2O$) and bischofite ($MgCl_2.6H_2O$) (BRAITSCH 1962).

Halite starts to precipitate after 90.9% of the original seawater volume has evaporated, whereas carnallite and bischofite precipitate when 99.2% and 99.4%, respectively, of the water has evaporated. More relevant to the process of chlorine isotope fractionation is the proportion of chloride that is precipitated from the evaporating seawater. Thus 82.5% of the original chloride content is precipitated as halite. Next, kainite starts to form, and, after 86.9% of the chloride is removed from the brine, carnallite begins to precipitate. Finally, when 88.9% is removed, bischofite crystallizes (BRAITSCH 1962).

-Rayleigh fractionation model of $\delta^{37}Cl$ in evaporites

The change in $\delta^{37}Cl$ caused by the precipitation of salts from evaporating seawater was modeled following a similar approach as HOLSER & KAPLAN (1966) used in their model of sulphur isotope variations in sulphates. The assumption that fractionation factors for all formed minerals is the same was ommitted. This model was derived from McINTIRE (1963) to calculate trace element partition coefficients in systems such as crytallizing melts.

The fractionation factor α is defined as:

$$\alpha = \frac{(^{37}Cl/^{35}Cl)_{precipitate}}{(^{37}Cl/^{35}Cl)_{solution}} \tag{1}$$

Assuming ideal Rayleigh behaviour, the $^{37}Cl/^{35}Cl$ ratio of the chloride in the solution changes according to:

$$\frac{r_c}{r_0} = \exp\left([\alpha-1]\ln\left[1-\frac{m_c}{m_0}\right]\right) \tag{2}$$

where r_c/r_0 is the ratio between $^{37}Cl/^{35}Cl$ in the sample and $^{37}Cl/^{35}Cl$ in the initial seawater, and m_c/m_0 is the weight fraction of chloride precipitated relative to the initial amount available.

The $\delta^{37}Cl$ of the brine, from the moment the first halite is precipitated can be calculated as:

$$\delta^{37}Cl = 1000 * \left(\frac{r_c-r_0}{r_0}\right) \tag{3}$$

and of the precipitate by factoring in $[\alpha-1]$:

$$\delta^{37}Cl = 1000 * \left(\alpha-1+\frac{r_c-r_0}{r_0}\right) \tag{4}$$

For each of the four chloride minerals (halite, kainite, carnallite and bischofite) a different fractionation factor must be used, depending on composition. From our experimentally determined fractionations for NaCl, KCl and $MgCl_2.6H_2O$ we calculated the fractionation factors for these minerals, see **table 1**.

Using these fractionation factors, the isotopic evolution of the evaporating brine and the salts that precipitate from it were calculated. The results are given in FIG. 1.

Table 1: Fractionation factors for halite, kainite, carnallite and bischofite, based on experimentally determined fractionation factors for NaCl, KCl and $MgCl_2.6H_2O$.

Mineral	Composition[1]	Fractionation factor
halite	100% NaCl	1.000235±0.000073
kainite	38.2% NaCl, 20.6% KCl, 41.2% $MgCl_2.6H_2O$	1.000052±0.000085
carnallite	30.0% NaCl, 23.6% KCl, 46.4% $MgCl_2.6H_2O$	1.000028±0.000087
bischofite	0.5% NaCl, 0.2% KCl, 99.3% $MgCl_2.6H_2O$	0.999935±0.000095

[1]The percentages give the composition of the mineral phase after Braitsch (1962)

During the precipitation of halite, $\delta^{37}Cl$ of the brine decreases from 0.00 to -0.42‰. Although kainite and carnalite have much smaller fractionation factors than halite, the total effect is still slightly higher than 1. Therefore, the $\delta^{37}Cl$ curve of the brine becomes near horizontal. When finally bischofite precipitates, which has a fractionation factor smaller than 1, $\delta^{37}Cl$ of the brine starts to increase.

The isotopic compositions of the salts that precipitate from the brine are also shown in FIG. 1. They deviate by $10^3 \ln\alpha$ from the curve for $\delta^{37}Cl$ of the brine. Because the fractionation factor is different for each salt mineral, a discontinuity occurs in $\delta^{37}Cl$ of the precipitate each time when a new salt mineral starts to precipitate.

Fig. 1: Calculated $\delta^{37}Cl$ of the precipitate and the remaining brine.

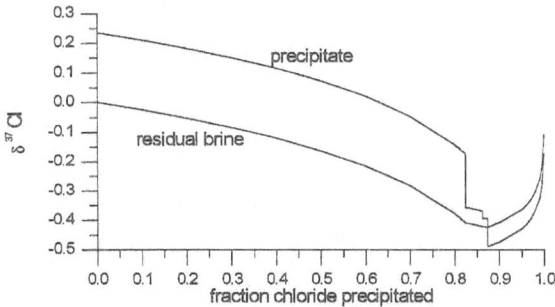

-Discussion of the $\delta^{37}Cl$ evolution model

The results in FIG. 1 predict that during the halite stage the $\delta^{37}Cl$ changes monotonically. This suggests that it is possible to use chlorine isotopes to estimate the amount of evaporation that has taken place. At later stages, where K- and Mg-salts precipitate, the method will be less effective since discontinuities appear and the fractionation curve is nearly flat. A remarkable result is the reversal of the fractionation as soon as bischofite precipitates. This indicates that in evaporites $\delta^{37}Cl$ has a theoretical minimum value of -0.49‰. Theoretically, $\delta^{37}Cl$ can go to high positive values during the very last stages of bischofite precipitation. It is not likely, however, that this will happen in nature because at this stage nearly all the water has disappeared and the remaining water co-precipitates in the very hygroscopic bischofite.

Finally it must be noted that the model assumes Rayleigh fractionation, i.e. no isotope exchange after the salt has precipitated. It is likely that in natural deposits some exchanges will take place, and that the Rayleigh fractionation model is not absolutely adhered to.

CASE STUDY: SALT FROM A ZECHSTEIN CORE

-Sample location and geologic setting

The isotope effects predicted by our model were compared with a series of samples from Zechstein core TR-2 (Veendam (near Groningen), The Netherlands, location WHC-2), made available by Billiton Refractories B.V.). The salt core, from the Veendam structure (FIG. 3), was drilled for magnesium exploration (e.g COELEWIJ *et al.* 1978, BUYZE & LORENZEN 1986). It contains carnallite and bischofite in addition to kainite and halite. This allows the study of the final stages of the evaporation process as well.

The Zechstein (Upper Permian) is the period during which the largest salt deposits in the history of the earth were deposited (BRAITSCH 1962). The core we studied is from the upper part of the Zechstein III formation (N.A.M & R.G.D. 1980). It samples the sub-formations Zechstein III-1, -2 and -3. All three sub-formations are again sub-divided in a lower part *a*, which contains mainly halite, and a carnallite- and bischofite-rich upper part *b*.

Twenty five samples were taken from the core between 1628 to 1786 meter depth. Their stratigraphic position is shown in FIG. 3. From 1628 to 1640 meter it contains halite with some carnallite (Zechstein III-3a). Between 1640 and 1654 meter carnallite, halite, and kieserite are present. Here the core is red with some banding (Z-III-2b). From 1654 to 1693 Z-III-2a, halite with some carnallite occur. Below this, a layered sequence (Z-III-1a, 1693-1780 m) of carnallite, bischofite, halite, and some kieserite is found. The lowest section sampled contains sylvite (KCl) and langbeinite ($K_2Mg_2(SO_4)_3$). Bolow this, massive halite is found (Z-III-1a, HAUG 1982). The two potassium-magnesium cycles, Z-III-1b and -2b are very different (COELEWIJ *et al.* 1978). Z-III-2b shows no distinct subcycles, whereas in Z-III-1b nine subcycles are present, each containing several mini-cycles. Of Z-III-2 a sample was taken every 10 meters and analysed for $\delta^{37}Cl$. Because of the more irregular behaviour of Z-III-1 sampling was less uniform because of the irregular alternation of halite rich and halite poor layers.

Table 2: *Mineralogical composition and results of δ³⁷Cl measurements of the salt samples.*

depth (m)	δ³⁷Cl (‰ vs. SMOC)	carnallite (%)	bischofite (%)	halite (%)	kieserite (%)
1628	-0.50±0.12	46	27	27	-
1638	-0.32±0.04	16	2	82	-
1648	-0.46±0.03	41	8	51	-
1658	-0.37±0.02	22	2	75	-
1668	-0.16±0.02	6	3	90	-
1678	-0.10±0.03	5	4	91	-
1688	-0.04±0.16	5	4	91	-
1692	-0.05±0.10	9	2	89	-
1710	+0.24±0.06	4	2	91	2
1716	-0.16±0.08	42	8	39	11
1718	-0.22±0.14	16	42	19	6
1725	-0.18±0.04	15	50	25	10
1736	-0.34±0.07	45	2	25	28
1738	-0.56±0.09	45	2	25	28
1740	-0.33±0.02	45	2	25	28
1742	-0.42±0.01	47	-	52	1
1744	-0.23±0.05	5	-	93	1
1747	-0.25±0.01	62	-	33	4
1751	-0.43±0.11	62	-	33	4
1756	-0.47±0.06	57	-	32	7
1758	-0.26±0.03	8	-	90	2
1771	-0.14±0.10	-	-	93	5
1775	-0.22±0.01	-	-	95	3
1783	-0.58±0.04	-	-	45	25
1786	-0.09±0.04	-	-	97	2

-Results

δ³⁷Cl data, depths and salt compositions of the samples are shown in **table 2** and in FIG. 3. A distinction can be made between the halite stages (Z-III-1a, Z-III-2a, Z-III-3a) and the more advanced evaporite stages (Z-III-1b, Z-III-2b, Z-III-3b). Samples from the three halite stages have a simple salt chemistry, although several percent of carnallite, bischofite and kieserite may be present. δ³⁷Cl generally decreases in the halite stage samples in upward direction. Samples from the three advanced evaporite stages have high

Fig. 3: Stratigraphic column of drillhole TR-2 (left) and results of $\delta^{37}Cl$ measurements (right)

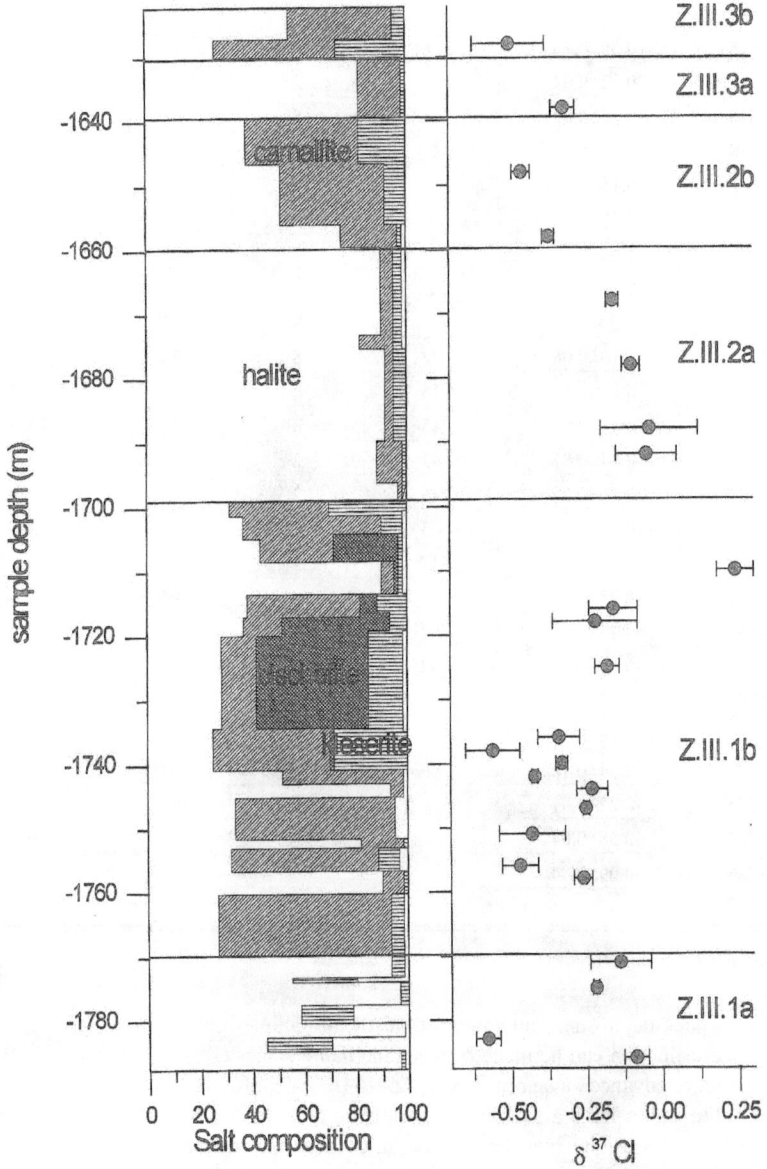

carnalite, bischofite or kieserite contents. Their δ^{37}Cl is generally at the lower end of the scale. The advanced stage Z-III-1b shows a complex pattern with δ^{37}Cl variations that seem to be partly related to the salt chemistry; increases in δ^{37}Cl correlate with increases in halite content. The two bischofite-rich samples in Z-III-1b have relatively high δ^{37}Cl values, despite the fact that they are the most evaporated samples in the section.

DISCUSSION

According to Coeleweij *et al.* (1978), the sub-formation Z-III-2 (upper half of the core) represents one single evaporation cycle. We, therefore, used this section to test our Rayleigh crystallization model. The sub-formation consists of two distinctive parts. The lower one (Z-III-2a) contains mainly halite, whereas the upper part (Z-III-2b) represents the more advanced stages of evaporation and has a complex salt chemistry. As predicted by the Rayleigh crystallization model, δ^{37}Cl decreases gradually as evaporation proceeds.

Sub-formation Z-III-3 is represented by only two samples, one at the onset of Z-III-3a and one at the end of Z-III-3a. Also here δ^{37}Cl at the end of Z-III-2a is lower than at the beginning, indicating that these samples are related by an evaporation process. However, the lower sample (1638), has a δ^{37}Cl of -0.32‰, much lower than would be expected at an early stage in the evaporation cycle. This suggests that an influx of seawater caused substantial dissolution of δ^{37}Cl salt, so that the chlorine isotope composition of the brine became relatively light. The composition of the two Z-III-3a samples, which contain substantial carnallite contents, supports this model.

The lower half of the core was taken from the Zechstein III-1 sub-formation which is much more complicated. According to Coeleweij *et al.* (1978), this part of the core represents nine evaporation cycles with several sub-cycles. 17 samples were taken from this section. Obviously, the sampling cannot be representative of this section. Several of the samples are not even from the same cycle. For example, consider the two lowest samples from Z-III-1a, sample 1789 is nearly pure halite and has a δ^{37}Cl of -0.09‰, which is quite normal. Sample 1783, however, has the most negative δ^{37}Cl found in this study. It contains abundant langbeinite and sylvite, which is indicative of secondary processes (Braitsch 1962), that may have changed the isotopic composition. The overlying Zechstein III-1b is an alternation of layers dominated by halite, carnallite and bischofite. In this section, δ^{37}Cl is generally low, as expected in the advanced stages of evaporation.

In these sections, halite-rich layers tend to have a higher δ^{37}Cl than the adjacent carnallite-rich layers, which is compatible with an influx of seawater. In most cases, however, δ^{37}Cl values remain relatively low, suggesting that substantial amounts of the earlier salts were dissolved when the fresh seawater entered the basin. The uppermost Z-III-1b sample (1710) is almost pure halite with a very high δ^{37}Cl, +0.24‰. This sample must have been crystallized from an influx of fresh seawater inflow and represents almost the first precipitate. Particularly interesting are the two bischofite-rich layers in sub-formation Z-III-1b. The layers have a relatively high δ^{37}Cl compared to other samples from similarly advanced evaporation stages. This supports the conclusion reached in the experimental section, that fractionation reverses as soon as bischofite precipitates and that $\delta^{37}Cl$ increases during the last stages of evaporite formation.

CONCLUSIONS

The $\delta^{37}Cl$ results on Zechstein evaporites and the $\delta^{37}Cl$ behaviour predicted by a Rayleigh fractionation model are in reasonable agreement. Crucial to our $\delta^{37}Cl$ model are the significantly different, experimentally determined fractionation factors for the different salts. These data are contrary to predictions made in early (chlorine) isotope studies (HOERING & PARKER 1961, HOLSER & KAPLAN 1966), where it was assumed that fractionation factors are similar for all salt compositions. An implication of the changes in isotope fractionation factor is that highly negative $\delta^{37}Cl$ values cannot be generated in the final stages of evaporite formation. This finding contradicts predictions made in earlier chlorine isotope studies of evaporites. In these studies these negative values have been looked for but were never found.

The systematic changes in $\delta^{37}Cl$, especially during the halite stage, are a useful tool in the study of evaporites. They can be used to monitor the amount of evaporation that has taken place, detect periods of input of new seawater, and give information on mixing with partly redissolved salt.

ACKNOWLEDGMENTS

We thank Billiton Refractories B.V., and especially ing. H. Lorenzen and ing. H.P. Rogaar for providing the samples and information about the samples. S.O. Scholten carefully read an earlier version of the manuscript and suggested many improvements. Miss D.C. McCartney is thanked for linguistic advice. The mass spectrometer was partly financed by the Netherlands Organization for the Advancement of Science (NWO). This research is part of the Awon project "Geochemistry of chlorine isotopes" (№ 751.355.014).

REFERENCES

AULT W.U. & KULP J.L. (1959) Isotopic geochemistry of sulphur. *Geochim. Cosmochim. Acta* **16** 201-235

BRAITSCH O. (1962) *Entstehung und stoffbestand der Salzlagerstätten.* Springer verlag. 232 pp.

BUYZE D. & LORENZEN H. (1986) Solution mining of multi-component magnesium-bearing salts - a realization in the Netherlands. *CIM bull.* **79 (889)** 52-60

COELEWIJ P.A.J., HAUG G.M.W. & VAN KUIJK H. (1978) Magnesium-salt exploration in the Northeastern Netherlands. *Geol. Mijnb.* **57** 487-502

ERIKSSON E. (1963) The yearly circulation of sulfur in nature. *J. Geophys. Res.* **68** 4001-4008

HAUG G. (1982) *Report on the results of drillhole TR-2.* Internal Shell Report.

HOERING T.C. & PARKER P.L. (1961) The geochemistry of the stable isotopes of chlorine. *Geochim. Cosmochim. Acta* **23** 186-199

HOLSER W.T. & KAPLAN I.R. (1966) Isotope geochemistry of sedimentary sulfates. *Chem. Geol.* **1** 93-135

KAUFMANN R.S. (1984) *Chlorine in groundwater: stable isotope distribution.* Ph.D. thesis. University of Arizona. 137 pp.

KAUFMANN R.S., LONG A., BENTLEY H. & DAVIS S. (1984) Natural chlorine isotope variations. *Nature* **309** 338-340

KAUFMANN R.S., LONG A. & CAMPBELL D.J. (1988) Chlorine isotope distribution in formation waters, Texas and Louisiana. *AAPG Bull.* **72** 839-844

MCINTAIRE W.L. (1963) Tace element partition coefficients - a review of theory and applications to geology. *Geochim. Cosmochim. Acta* **27** 1209-1264

NEDERLANDSE AARDOLIE MAATSCHAPPIJ B.V. & RIJKS GEOLOGISCHE DIENST (1980) Stratigraphic nomenclature of the Netherlands. *Verh. Kon. Ned. Geol. Mijnb. Gen.* **32** 77 pp.

NIELSEN H. (1966) Schwefelisotope im marinen Kreislauf und das δ^{34}S der früheren Meere. *Geol. Rundschau* **55** 160-172

NIELSEN H. & RICKE W. (1966) Schwefel-Isotopenverhaltnisse von Evaporiten aus Deutschland; Ein Beitrag zur Kenntnis von δ^{34}S im Meerwasser-Sulfat. *Geochim. Cosmochim. Acta* **28** 577-591

REES C.E. (1970) The sulphur isotope balance of the ocean: an improved model. *Earth Planet. Sci. Lett.* **7** 366-370

TAYLOR J.W. & GRIMSRUD E.P. (1969) Chlorine isotopic ratios by negative ion mass spectrometry. *Anal. Chem.* **41** 805-810

THODE H.G. & MONSTER J. (1965) Sulphur-isotope geochemistry of petroleum, evaporites and ancient seas. *AAPG Bull., Mem.* **4** 367-377

VENGOSH A., CHIVAS A.R. & MCCULLOCH M.T. (1989) Direct determination of boron and chlorine isotopic compositions in geological materials by negative thermal-ionization mass spectrometry. *Chem. Geol. (Isot. Geosci. Sect.)* **79** 333-343

CHAPTER 11

Stable Chlorine Isotopes in Rocks
a New Method For Extraction
Chlorine From Rocks
Case Study: the Ilímaussaq Intrusion, South Greenland

H.G.M. Eggenkamp[1]

ABSTRACT-- A new method was developed to extract chlorine from rocks for $\delta^{37}Cl$ measurements. For this method the sample is heated with NaOH, and this is dissolved in water. After acidifying and removal of silica, chloride ions are precipitated as AgCl. The method is used to measure $\delta^{37}Cl$ in rock samples from the Ilímaussaq intrusion in south Greenland. The observed $\delta^{37}Cl$ variations were small (-0.02 to +0.32‰), most probably due to the high crystallization temperatures of the rocks.

INTRODUCTION

In most earlier studies on chloride isotope geochemistry the isotopic variation was measured on aqueous solutions or on substances that are soluble in water. Only a few papers include data on silicate rocks (HOERING & PARKER 1961, OWEN & SCHAEFFER 1954). These earlier data on silicate rocks did not show significant isotopic variations, which is partly due to the large errors of the old mass spectrometers. Since modern techniques can measure much more accurately, chlorine isotope variations in silicate rocks can now be studied. We selected samples from the Ilímaussaq intrusion (Greenland) because the chlorine content in these samples is high, and they can be measured relatively easily. A new, relatively simple method has been developed to extract the chloride from silicate samples.

MEASUREMENT OF δ^{37}CL IN ROCK SAMPLES

-Preparation of samples for isotope measurement

A major problem with silicate rock samples is their low chlorine content. In order to prevent contamination, the amounts of solutions and other chemicals must be minimized. This is also important at the stage where Cl⁻ is precipitated as AgCl.

A method was developed which is partly derived from the NaOH method used by BEHNE (1953). The procedure is as follows:

X grams of powdered rock, in which X is an amount of rock containing enough chloride to perform one or more $\delta^{37}Cl$ measurements, is heated together with 10X grams of NaOH pellets for about 30 minutes in a nickel crucible. The rock dissolves in the

1Department of Geochemistry, Utrecht University, P.O.Box 80.021, 3508 TA Utrecht, The Netherlands

molten NaOH and Si-O bonds are partly destroyed. The temperature must not be too high since NaOH will evaporate at high temperatures (e.g vapour pressure is 1 mmHg (1.33 mbar) at 739 °C and 10 mmHg (13.3 mbar) at 897 °C (STULL 1947)). After cooling, the sample is dissolved in 35X ml H_2O (bidest). This is achieved by putting the nickel crucible in a beaker with water on a magnetic stirrer. After two hours of stirring the content of the crucible is dissolved or suspended in the water.

This suspension must be exposed to air for some time (overnight) to oxidize Fe^{2+} and other ions. Because the solution has an extremely high pH, which would cause precipitation of Ag_2O after addition of Ag^+ ions, the solution must be acidified. This is done by adding 17.5X ml HNO_3 65% which produces a colloidal "solution" of silica gel according to:

$$SiO_4^{4-} + 4HNO_3 \rightarrow H_4SiO_4 \downarrow + 4NO_3^- \qquad (1)$$

This colloidal solution can not be filtered and, therefore, 3X ml HF 40% is added. The silica gel reacts with the HF to form a combined silica oxyfluoride.

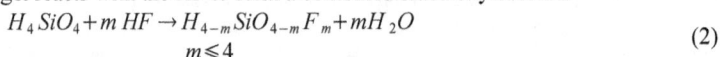

$$H_4SiO_4 + m\,HF \rightarrow H_{4-m}SiO_{4-m}F_m + mH_2O \qquad (2)$$
$$m \leqslant 4$$

This reaction does not go to completion. The oxyfluoride precipitates slowly (overnight) to the bottom of the glass container. Then 10.9X ml of a 2.5 mol.kg^{-1} $Mg(NO_3)_2.6H_2O$ solution is added (64.1 gram $Mg(NO_3)_2.6H_2O$ and 35.9 gram H_2O, density 1.25 gr.ml^{-1}). Excess toxic HF will react to poorly soluble MgF_2.

The resulting suspension is filtered. The residue contains the reacted silicagel, the MgF_2, the insoluble hydroxides and rock components that were not dissolved. Cl$^-$ is concentrated in the filtrate.

The filtrate has a low pH and a high ionic strength, so that the silver chloride can be precipitated directly from this solution without adding a buffer or KNO_3.

-Testing

The method was tested by analysing samples of schist and gabbro, both with very low chloride contents, to which seawater or known quantities of NaCl ware added. Also blanks, consisting of seasand (Merck®) and pure NaOH were made. These blanks gave CH_3Cl yields which were too low to be measured. Besides, it was checked that no fractionation occurs upon precipitating AgCl from the very acid and very concentrated solution.

Fractionation upon precipitating AgCl was tested using NaCl from Merck®. A 12000 ppm Cl. solution was made, which was measured for $\delta^{37}Cl$. From this solution an acid high ionic strength solution was made by mixing 25 ml of the 12000 ppm solution with 25 ml 8 N NaOH and 50 ml HNO_3 65 %. Within the analytical error the $\delta^{37}Cl$ results of the two solutions were equal, -0.05±0.09‰ for the normal solution and -0.04±0.02‰ for the acidified solution.

Seawater or NaCl with a known $\delta^{37}Cl$ value was added to two rock samples which very low chloride contents. The first sample was a schist, HH98 from the Beira-schist complex, Portugal (RIBEIRO 1974). This sample consists mainly of quartz and muscovite

---

Fig. 1: Geological map of the Illimaussaq alkaline intrusion (from BAILEY et al. 1981).

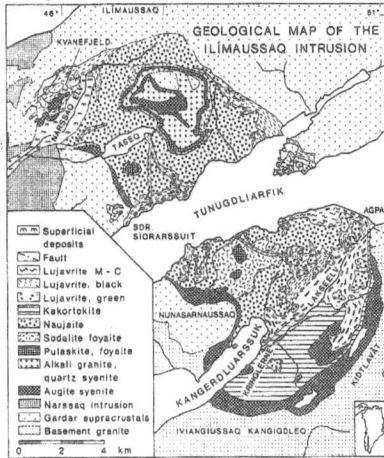

(both slightly less than 50 %), some Fe-oxides (about 5%), tourmaline, biotite, zircon and apatite (all less then 1 %, HOLLANDERS & DE HAAS 1986). 1 ml seawater per gram sample, or 10 % of NaCl was added to the sample. The average $\delta^{37}Cl$ value of the sample with added seawater is +0.14±0.05‰, the average $\delta^{37}Cl$ of the sample with added NaCl is +0.09±0.08‰. The second sample was a Norwegian gabbro sample obtained from DAM (1994). To this sample was also added 10% NaCl. The resulting $\delta^{37}Cl$ deviated remarkably from the added NaCl; the average value is +0.39±0.10‰. The reason for this systematic error is not known.

THE ILÍMAUSSAQ INTRUSION

To test the method, the Ilímaussaq intrusion of Southern Greenland was studied. This highly alkaline intrusion was selected because it is relatively rich in chlorine. The intrusion, a part of the Gardar province, is about 17 by 8 km (e.g. USSING 1912, FERGUSON 1964, SØRENSEN 1967, BAILEY et al. 1981). Its Rb-Sr isotope age is 1168±21 Ma (BLAXLAND et al. 1976). The intrusion was emplaced in three main pulses; an augite syenite stage, an alkali acid rock stage, and a final agpaitic stage. The last stage is the most important; it is divided in four substages, a pulaskite-foyaite substage, a sodalite foyaite-naujaite substage, a kakortokite substage, and a lujavrite substage. (See FIG. 1 for a geologic map.) The intrusion is agpaitic, which is defined as a peralkaline nepheline syenites containing complex volatile-bearing minerals such as eudialite and rinkite (SØRENSEN 1960, 1974). In these rocks Cl-rich minerals are common. Fluid inclusions (KONNERUP-MADSEN & ROSE-HANSEN 1982, 1984, KONNERUP-MADSEN et al. 1979, 1981) and stable isotopes in these inclusions (KONNERUP-MADSEN et al. 1988) have been studied extensively.

In the Ilímaussaq intrusion, fractional crystallization and gas pressure changes (e.g. pH_2O, $pHCl$) played a major role (SØRENSEN 1969). During crystallization chlorine is preferentially partitioned in the gaseous or aqueous phase (CARMICHEL *et al.* 1974). The highest chlorine contents are found in the sodalite bearing rock types of which naujaite is the most important, and sodalite-foyaite the second important. Theoretically, isotope fractionation depends on temperature, and at the high temperatures during the intrusion, fractionation should be small (KAUFMANN 1989).

-Sample material

Six rock samples from the intrusion have been measured for $\delta^{37}Cl$. The samples are from different substages within the agpaitic stage, in order of increasing differentiation, 154350 (a sodalite foyalite), 154309 (a naujaite), H-1 and H-2 (kakortokites), 57033 and 154367 (lujavrites, see **table 1**). Chloride concentrations are the highest in the naujaite (about 1.9%) and the lowest in the lujavrites (about 0.1%). The chloride contents found for the lujavrite and the kakortokite are in agreement with the values measured by GERASIMOVSKI & TUZOVA (1964).

-Results

Results are shown in **table 1** and FIG. 2. Four samples have a similar $\delta^{37}Cl$ of about +0.10‰, one sample (H-1) is slightly negative. One sample (154367) has a significantly different $\delta^{37}Cl$ value of +0.32‰.

Table 1: *Chloride content and $\delta^{37}Cl$ of the Ilímaussaq rock samples.*

Sample code	Cl⁻ (%)	$\delta^{37}Cl$ (‰)	Stage
154350	0.39	0.09	sodalite-foyalite
154309	1.92	0.11	naujaite
H-1	0.21	-0.02	kakortokite
H-2	0.35	0.09	kakortokite
57033	0.11	0.11	lujavrite
154367	0.06	0.32	lujavrite

-Discussion

With the exception of sample 154367 the rocks have similar $\delta^{37}Cl$ values. Equilibrium fractionation generally decreases with increasing temperatures. See e.g. BOTTINGA (1969) for fractionation curves of carbon isotopes and OHMOTO & RYE (1979) for

Fig. 2: Relationship between Cl⁻ content and δ^{37}Cl in measured Illimaussaq rocks.

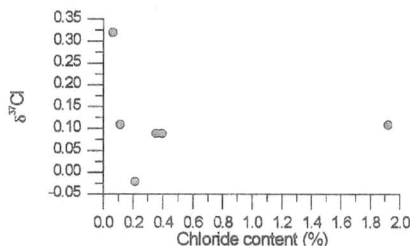

sulphur isotopes. KAUFMANN (1989) calculated fractionation factors for Cl_2/NaCl and HCl/NaCl equilibria. He used δ^{37}Cl as a geothermometer. The temperature was calculated from the difference of the maximum and the minimum δ^{37}Cl in the system. Comparing KAUFMANN's (1989) results with our measurements (maximum difference is 0.34‰), the temperature of the substage should be higher than 800 °C. It is also possible that during crystallization at high temperatures the isotope fractionation is much smaller than predicted by KAUFMANN's (1989) data. EGGENKAMP et al. (1993) concluded that the fractionation factor between dissolved and solid chloride in precipitating salt minerals at about 20 °C is 0.24‰ or less. If the fractionation in rock forming minerals at high temperatures is much smaller, differences will be too small to determine, except when Rayleigh fractionation has taken place.

Crystallization temperatures in agpaitic rocks strongly depend on the water pressure in the magma. They were estimated for lujavrite by KOGARKO & ROMANCHEV (1978) to be around 600 to 700 °C at 1 kB water pressure (which presumably is the pressure in Ilímaussaq (SØRENSEN 1969)). LARSEN (1976) concluded that the Ilímaussaq rocks had an extraordinary long crystallization interval, ranging from 800-900 °C to about 500 °C during the late stages of crystallization. This estimate for the initial temperature is in agreement with SØRENSEN (1969), while the low temperature in the late stages is in agreement with melting experiments of PIOTROWSKI & EDGAR (1970) and SOOD & EDGAR (1970).

KAUFMANN (1989) used δ^{37}Cl in water samples and not in rocks, so his data can not be compared directly to those presented here. The effect of the rock matrix is not yet understood. In the present study the measured δ^{37}Cl values seem to be not in contradiction with the crystallization temperatures, since the δ^{37}Cl values only define a minimum temperature (i.e. about 800 °C).

-Conclusions

The δ^{37}Cl of six rock samples from the Ilímaussaq intrusion was measured. δ^{37}Cl shows no clear relation with the differentiation trend. The samples, except one have a slightly positive δ^{37}Cl value, which probably reflects equilibrium fractionation. The one sample that has a somewhat higher $\delta^{37}Cl$ value is from the last magmatic stage, but this is insufficient evidence that some fractionation occurred lower temperature.

A problem occurred with the test samples. The metamorphic schist samples gave satisfactory results. The gabbro sample, however, seemed to show a matrix effect. Also concentration effects are not examined. As the $\delta^{37}Cl$ values of the measured rock samples fall in a close range, the values seem to be acceptable.

ACKNOWLEDGEMENTS

J. Konnerup-Madsen and J. Rose-Hansen provided the Ilímaussaq samples from which $\delta^{37}Cl$ measurements are presented. M.E.L. Kohnen measured chloride contents of the samples. R. Kreulen and S.O. Scholten critically reviewed an earlier draft of this manuscript. This research is sponsored by the Netherlands Organization for the Advancement of Science (AWON grant 751.355.014)

REFERENCES

BAILEY J.C., LARSEN L.M. & SØRENSEN H. (1981) Introduction to the Ilímaussaq intrusion with a summary of the reported investigations. *Rapp. Grønlands Geol. Unders.* **103** 5-17

BEHNE W. (1953) Untersuchungen zur Geochemie des Chlor und Brom. *Geochim. Cosmochim. Acta* **3** 186-214

BLAXLAND A.B., VAN BREEMEN O. & STEENFELT A. (1976) Age and origin of agpaitic magmatism at Ilímaussaq, south Greenland: Rb-Sr study *Lithos* **9** 31-38

BOTTINGA Y. (1969) Calculated fractionation factors for carbon and hydrogen isotope exchange in the system calcite-carbon dioxide-graphite-methane-hydrogen-water-vapor. *Geochim. Cosmochim. Acta* **33** 49-64

CARMICHAEL I.S.E., TURNER & VERHOOGEN (1974) *Igneous Petrology* McGraw Hill Book Co., New York. 739 pp.

DAM B.P. (1994) *Ph. D. Thesis*, in prep.

EGGENKAMP H.G.M., KREULEN R. & KOSTER VAN GROOS A.F. (1993) Fractionation of chlorine isotopes in evaporites. *Terra Abs.* **5 (1)** 650

FERGUSON J. (1964) Geology of the Ilímaussaq alkaline intrusion, South Greenland. Part I. Description of map and structure. *Bull. Grønlands Geol. Unders.* **39** (also *Meddr Grønland* **172**,4) 81 pp.

GERASIMOVSKY V.I. & TUZOVA A.M. (1964) Geochemistry of chlorine in nepheline syenites. *Geochem. Int.* **1964** 855-867

HOERING T.C. & PARKER P.L. (1961) The geochemistry of the stable isotopes of chlorine. *Geochim. Cosmochim. Acta* **23** 186-199

HOLLANDERS M.A. & DE HAAS G.J.L.M. (1986) *The petrology and geochemistry of the granite-schist complex of the Viseu region, Northern Portugal.* Unpubl. M.Sc. Thesis, University of Utrecht. 60 pp.

KAUFMANN R.S. (1989) Equilibrium exchange models for chlorine stable isotope fractionation in high temperature environments. *Proc. WRI* **6** 365-368

KOGARKO L.N. & ROMANCHEV B.P. (1978) Temperature, pressure, redox conditions, and mineral equilibria in agpaitic nepheline syenites and apatite-nepheline rocks. *Geochem. Int.* **14** 113-128

KONNERUP-MADSEN J. & ROSE-HANSEN J. (1982) Volatiles associated with alkaline igneous rift activity: fluid inclusions in the Ilímaussaq intrusion and the Gardar granitic complexes (South-Greenland). *Chem. Geol.* **37** 79-93

KONNERUP-MADSEN J. & ROSE-HANSEN J. (1984) Composition and significance of fluid inclusions in the Ilímaussaq peralkaline granite. South Greenland. *Bull. Minéral.* **107** 317-326

KONNERUP-MADSEN J., LARSEN E. & ROSE-HANSEN J. (1979) Hydrocarbon-rich inclusions in minerals from the alkaline Ilímaussaq intrusion. *Bull. Minéral.* **102** 642-653

KONNERUP-MADSEN J., ROSE-HANSEN J. & LARSEN E. (1981) Hydrocarbon gases associated with alkaline igneous activity: evidence from compositions of fluid inclusions. *Rapp. Grønlands Geol. Unders.* **103** 99-108

KONNERUP-MADSEN J., KREULEN R. & ROSE-HANSEN J. (1988) Stable isotope characteristics of hydrocarbon gases in the alkaline Ilímaussaq complex, south Greenland *Bul. Minéral.* **111** 567-576

LARSEN L.M. (1976) Clinopyroxenes and coexisting mafic minerals from the alkaline Ilímaussaq intrusion, South Greenland. *J. Petrol.* **17** 258-290

OHMOTO H. & RYE (1979) Isotopes of sulphur and carbon. *In: Geochemistry of hydrothermal ore deposits*, 2nd edition. Holt Rinehart and Winston, New York.

OWEN H.R. & SCHAEFFER O.A. (1954) The isotope abundances of chlorine from various sources. *J. Amer. Chem. Soc.* **77** 898-899

PIOTROWSKI J.M. & EDGAR A.D. (1970) Melting relations of undersaturated alkaline rocks from South Greenland. *Meddr. Grønland* **181**,9 1-62

RIBEIRO A. (1974) Contribution à l'étude tectonique de Tras-os-Montes oriental. *Ser. Geol. Port.* **Mem. 24, N. Ser. Atlas.** Lisbon, 177 pp.

SOOD M.K. & EDGAR A.D. (1970) Melting relations of undersaturated alkaline rocks from the Ilímaussaq intrusion and Grønnedal-Ika complex, South Greenland, under water vapour and controlled partial oxygen pressures. *Meddr Grønland* **181**,12 1-41

SØRENSEN H. (1960) On the agpaitic rocks. *Rep. 21nd Int. Geol. Congr. Norden* **13** 319-327

SØRENSEN H. (1967) On the history of exploration of the Ilímaussaq alkaline intrusion, South Greenland. *Bull. Grønlands Geol. Unders.* **68** (also *Meddr Grønland* **181**,4) 33 pp.

SØRENSEN H. (1969) Rhythmic igneous layering in peralkaline intrusions. An essay review on Ilímaussaq (Greenland) and Lovozero (Kola, USSR). *Lithos* **2** 261-283

SØRENSEN H. (1974) Alkali syenites, feldspathoidal syenites and related lavas. *In* SØRENSEN H. (edt.) *The alkaline rocks* 22-52. John Wiley & Sons, London.

STULL D.R. (1947) Vapour pressure of pure substances. Inorganic compounds. *Ind. Eng. Chem.* **39** 540-550

USSING N.V. (1912) Geology of the country around Julianehaab, Greenland. *Meddr Grønland* **38** 376 pp.

δ^{37}Cl Variations in Minerals

H.G.M. Eggenkamp[1] and R.D. Schuiling[1]

ABSTRACT-- δ^{37}Cl values are determined on minerals from different places and methods of formation. Evaporite minerals, secondary sedimentary minerals and magmatic minerals don't seem to have large δ^{37}Cl variations. In the first two this is caused by the limited fractionation factors in brine-precipitation systems, in the latter it is caused by the low fractionation factors in high temperature environments. Large variations are found in fumarole minerals and basic metal chloride minerals. These minerals probably form through escape of gaseous species. Fumarole minerals are formed of the gases and are found to have negative δ^{37}Cl values. Through repeated sublimation and condensation values as low as -4.9‰ are found. Basic metal chlorides formed from the residues and are found to have positive δ^{37}Cl values. The most extreme value found is +6‰ for an atacamite sample. It is supposed that these highly positive values probably can be used as an exploration tool.

INTRODUCTION

Chlorine-bearing minerals can form in different ways and under different conditions. The fractionation of their chlorine isotopes is likely to depend on the mode of formation. For this reason a selection of chloride minerals was analyzed for their ^{37}Cl/^{35}Cl ratio.

Earlier attempts to measure chlorine isotope variations in minerals revealed no significant variations (CURIE 1921, OWEN & SHAEFFER 1954, HOERING & PARKER 1961). Partly, this was due to the limited accuracy of the methods at that time. Sample selection, however, also is an important factor because in the present study relatively large variations are found that could easily have been detected in the earlier studies. One of the minerals, apatite from Ødegårdens Verk (Norway) was measured before (GLEDITSCH & SAMDAHL 1922, DORENFELT 1923, MORTON & CATANZARO 1964). No significant δ^{37}Cl variations were found in these studies.

MATERIAL AND METHODS

Twenty different chloride minerals were measured. Most of these were obtained from the Nationaal Natuurhistorisch Museum (NNM) at Leiden, other were obtained from the Mineralogisch-Geologisch Instituut (MGI) collection at Utrecht University and from Mr. B.P. Dam. The minerals have different chemical properties, so that different methods were needed to dissolve them before δ^{37}Cl analyses could be made. Methods to dissolve the minerals, except for silicates and borates, are taken from HINTZE (1915). The analyzed minerals are listed below in systematic order (after HÖLZEL 1989); their origin, environment of formation and the method used to dissolve them are briefly indicated.

HALITE (NaCl): This is an evaporitic mineral. Halite is the first chloride mineral that

---

1Department of geochemistry, Utrecht University, P.O.Box 80.021, 3508 TA Utrecht, The Netherlands

precipitates from seawater and precipitation starts when about 90% of the original seawater has evaporated. Our sample is №150.1880 from MGI. This specimen is from Staßfurt, Saxony, Germany. Halite has a very high solubility (ranging from 35.7 gram per 100 cc at 0°C to 39.1 gram per 100 cc at 100°C), so that the mineral can simply be dissolved in water prior to $\delta^{37}Cl$ analysis.

SAL AMMONIAC (NH_4Cl): This mineral is found in fumaroles where volcanic gasses escape to the surface (e.g. Lacroix 1907, Vertacnik 1983). Specimen №74.1907 from MGI comes from the Etna volcano, Sicily, Italy and was formed during the 1892 eruption. The mineral is very soluble in water, with solubilities ranging from 29.7 gram per 100 cc at 0°C to 75.8 gram per 100 cc at 100°C. For $\delta^{37}Cl$ measurements sal ammoniac is dissolved in water.

CALOMEL (Hg_2Cl_2): This mineral forms through oxidation of native mercury with chloride bearing solutions after the simplified formula:

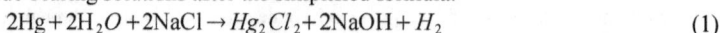

$$2Hg + 2H_2O + 2NaCl \rightarrow Hg_2Cl_2 + 2NaOH + H_2 \tag{1}$$

(Krupp *et al.* 1985). Our sample №RGM 26124 from NNM, comes from the Pfalz, Wolffstein, Germany. Calomel is barely soluble in water (0.2 mg per 100 cc at 25°C).

The following three methods were tried to get chloride ions in solution:
1) Reaction with KOH to form Hg_2O, H_2O and KCl.
2) Oxidation with $Hg(NO_3)_2$ to $HgCl_2$.
3) Dissolution in concentrated HNO_3.
All three methods failed. Although the mineral is soluble in $Hg(NO_3)_2$, it was not possible to precipitate silver chloride from this solution. Probably this is due to the covalent bonding of $HgCl_2$, keeping the Cl^- concentration extremely low.

CARNALLITE ($KMgCl_3.6H_2O$): This is an evaporitic mineral. It precipitates from seawater when about 98% of the water is evaporated. Sample №59^b.1896 from MGI comes from Leopoldhall-Staßfurt Anhalt, Germany. Carnallite is very soluble (64.5 gram per 100 cc at 19°C) and is dissolved in water prior to $\delta^{37}Cl$ analysis.

ATACAMITE ($Cu_2(OH)_3Cl$): This mineral can be formed in the oxidation zones of copper mineralizations, under neutral pH (about 7), high copper and high chloride concentrations. Atacamite is typically found in desert environments (Rodriguez 1980). Sample RGM 82272 from NNM, was found at Ravensthorpe, Australia. It is dissolved in dilute (1:5) nitric acid.

CONNELLITE ($Cu_{19}[Cl_4|(OH)_{32}|SO_4].3H_2O$): A mineral from oxidation zones of copper mineralizations. Our sample RGM 80684 from NNM was found in Cornwall, England. It is dissolved in dilute (1:5) nitric acid.

COTUNNITE ($PbCl_2$): Is formed in a similar way as sal ammoniac, in fumaroles (Lacroix 1907). The sample (RGM 82292 from NNM) is from the Vesuvius in Italy. Since the solubility is relatively high (0.99 gr per 100 cc at 20°C) it is dissolved in water.

PARALAURIONITE (RAFAELLITE, PbCl(OH)): The mineral is found in the oxidation zones of sulphide mineralizations and is stable when the chloride concentrations are sufficiently high (Humphreys *et al.* 1980). Sample RGM 5514 from NNM, was found in South Mendoza, Argentina. The mineral is dissolved in warm 10% nitric acid.

132

NADORITE ($PbSbO_2Cl$): Mineral from oxidation zones of lead-bismuth mineralizations. Our sample is № RGM 23923 from NNM, found in Djebel Nador, Algeria. It is dissolved in a mixture of tartaric acid and dilute nitric acid.

BOLEITE ($Ag_{10}Pb_{26}Cu_{24}Cl_{62}(OH)_{48}.3H_2O$): Oxidation product of complex sulphide ores. Sample is №RGM 79921 from NNM comes from the type locality at Boléo, Mexico. Dissolved in dilute (1:5) nitric acid prior to $\delta^{37}Cl$ analysis.

BORACITE ($Mg_3[Cl|B_7O_{13}]$): Boracite is formed as a secondary mineral in salt deposits (APOLLONOV *et al.* 1988, HEIDE *et al.* 1980, HODENBERG *et al.* 1987). Boron is concentrated in the residual solution. It is also possible that a boron rich solution is added later. The boracite probably forms at increased pressure and temperature (BRAITSCH 1962). Our sample №RGM 82338 from NNM, was found in Lüneberg, Germany. It is dissolved in molten NaOH (see chapter 11).

HYDROXYLAPATITE ($Ca_5[OH,Cl,F|(PO_4)_3]$) and CHLORAPATITE ($Ca_5[Cl,OH,F|(PO_4)_3]$): Samples from Ødegårdens Verk, South Norway (DAM 1994). This occurrence, discovered in 1872 (BRÖGGER & REUSCH 1875), is one of the largest in Norway. The minerals occur in hydrothermal veins in a gabbro. Samples from the same locality have been measured in earlier chlorine isotope studies. The minerals are dissolved in nitric acid.

PYROMORPHITE ($Pb_5[Cl|(PO_4)_3]$), MIMETITE ($Pb_5[Cl|(AsO_4)_3]$) and VANADINITE ($Pb_5[Cl|(VO_4)_3]$): Secondary minerals formed by oxidation of lead-sulphide mineralizations. It is supposed that these minerals are deposited by circulating hydrothermal water (FÖRTSCH 1970). Pyromorphite is formed with the commonly available phosphate-ion. Mimetite forms when arsenic (a common element in sulphide ores) is available in sufficient concentration. Vanadium, necessary for the formation of vanadinite, may be derived from the surrounding sediments (SMIRNOV 1954). All three samples were obtained from NNM, pyromorphite №RGM 163934, was found in Broken Hill, Australia, mimetite №RGM 163496 is from Tsumeb, Namibia and vanadinite №RGM 411570 was found in Midelt, Marocco. The samples are dissolved in dilute (about 1:5) nitric acid; the vanadinite sample was too small to be analyzed for $\delta^{37}Cl$.

EUDIALITE ($Na_6Ca_2Ce_2Zr_2[(OH,Cl)_4|Si_3O_9|Si_9O_{25}]$): This mineral occurs in alkali rich intrusions, formed as residual fractionation products of alkali-basalt magma (KRIVDICK & TKACHUK 1988). Our sample is №RGM 162919 from NNM, found in St. Hilaire, Canada. The sample is dissolved in molten NaOH (see chapter 11).

AMPHIBOLES (approximately (FERROAN) PARGASITE, $NaCa_2(Mg,Fe)_4Al[(OH,Cl)_2|Al_2Si_6O_{22}]$): Samples GA159 and GA342 are from the coronitic Jomasknutene Metagabbro, Bamble, Norway. These minerals crystallized most probably in the presence of an aqueous fluid phase. The chloride concentration depends on the chloride concentration in the fluid, the distortion of the Si_4O_{11}-chain and the crystallization temperature (DAM 1994). The samples are dissolved in molten NaOH (see chapter 11).

SODALITE ($Na_8[Cl_2|(AlSiO_4)_6]$): Sodalite is found in nepheline-syenites (DEER *et al.* 1966). It can also be formed through contact metamorphism of sodium- and chloride-rich limestones (CUITIÑO 1986). The sample is №RGM 162939 found at St. Hilaire, Canada. It

Table 1: *Results of $\delta^{37}Cl$ measurements on minerals.*

Name	Formula	$\delta^{37}Cl$	σ_{n-1}
	Evaporite Minerals		
Carnallite	$KMgCl_3.6H_2O$	-0.48	0.07
Halite	$NaCl$	-0.10	0.09
	Fumarole Minerals		
Sal Ammoniac	NH_4Cl	-4.88	0.08
Cotunnite	$PbCl_2$	-0.46	0.05
	Magmatic Minerals		
Amphibole2		-0.26	0.12
Eudialite	$Na_6Ca_2Ce_2Zr_2[(OH,Cl)_4|Si_3O_9|Si_9O_{25}]$	-0.23	0.08
Amphibole1		-0.06	0.02
Chlorapatite	$Ca_5[Cl,F,OH|(PO_4)_3]$	-0.04	0.18
Sodalite	$Na_8[Cl_2|(AlSiO_4)_6]$	0.13	0.06
Hydroxylapatite	$Ca_5[OH,F,Cl|(PO_4)_3]$	0.92	
	Oxidation Minerals		
Paralaurionite (Rafaellite)	$PbCl(OH)$	-0.05	0.03
Calomel	Hg_2Cl_2	?	
Vanadinite	$Pb_5[Cl|(VO_4)_3]$	?	
Mimetite	$Pb_5[Cl|(AsO_4)_3]$	0.08	?
Nadorite	$PbSbO_2Cl$	0.08	0.22
Pyromorphite	$Pb_5[Cl|(PO_4)_3]$	0.25	?
Conellite	$Cu_{19}[Cl_4|(OH)_{32}|SO_4].3H_2O$	1.12	0.16
Boleite	$Ag_{10}Pb_{26}Cu_{24}Cl_{62}(OH)_{48}.3H_2O$	1.66	0.32
Atacamite	$Cu_2(OH)_3Cl$	5.96	0.17
	Secondary Sedimentary Minerals		
Boracite	$Mg_3[Cl|B_7O_{13}]$	0.10	0.21

is dissolved in molten NaOH (see chapter 11).

RESULTS

Results of the $\delta^{37}Cl$ measurements are found in **table 1**. Although substantial variations in $\delta^{37}Cl$ are found, only one value is lower then -1‰ (-4.88, sal ammoniac) and one is higher than +2‰ (+5.96, atacamite). These values belong to the most extreme $\delta^{37}Cl$ values ever measured. As can be seen in the **table 1**, the largest values are found in volcanic fumarole products and in oxidation zone minerals. Samples in other geologic environments are generally close to SMOC.

DISCUSSION

-Evaporites

The behaviour of chlorine isotopes is discussed in detail in chapter 10. Systematic $\delta^{37}Cl$ variations do occur in evaporites, but they are not expected to be very large because the fractionations between potassium/magnesium salts and the chloride solutions which they precipitate, are much smaller than for halite. The variations are explained by fractional crystallization from a brine with an original $\delta^{37}Cl$ of 0‰. The halite sample is supposed to represent an advanced stage of halite precipitation (see chapter 10, Fig. 1). Carnallite has a negative $\delta^{37}Cl$, which is in line with the evaporation model discussed in chapter 10.

-Secondary sedimentary minerals

Boron is concentrated during the evaporation of a brine. The boron concentration in evaporating seawater increases from 0.0012% at the beginning of gypsum precipitation to 0.05% at the beginning of bischofite precipitation (BRAITSCH 1962). Evaporitic borate minerals will precipitate only in the latest stage of evaporation. Boracite, however, has a limited solubility in water and formes as a secondary mineral in evaporite deposits, probably at somewhat elevated temperature and pressure (HODENBERG *et al.* 1987). Both the boron and the chloride contained must be derived from the salt, and therefore $\delta^{37}Cl$ can be expected to be compatible with the late stages of precipitation. The $\delta^{37}Cl$ value of the boracite is in line with this (see also chapter 10).

-Minerals from igneous and metamorphic rocks

These minerals were formed at high temperatures and only small $\delta^{37}Cl$ variations are expected because fractionation factors between fluid and solid phases are small at high temperatures (KAUFMANN 1989). $\delta^{37}Cl$ values close to zero suggest that the average chlorine isotopic composition of the crust is not much different from seawater. The $\delta^{37}Cl$ values of the metamorphosed minerals are in the same range, indicating that during metamorphism no significant fractionation occurs. Only the hydroxylapatite has a different value. This sample, however, was too small and the analysis may be in error (there was not enough material for a duplicate analysis).

-Fumaroles

The two measured fumarole minerals have negative δ^{37}Cl values. These minerals are formed when volcanic gas compounds condense as a solid mineral in the vent. The minerals have a high vapour pressure under the prevailing conditions and they may have sublimated and condensed several times before they were collected. Due to kinetic isotope fractionation, the gaseous sublimation product probably will have a lower δ^{37}Cl than the solid residue. Repeated sublimation and condensation, and partial loss of material, can be expected to produce substantial isotope effects both in positive and negative direction.

-Oxidation zone minerals, basic chlorides

δ^{37}Cl values of these minerals range from about 0‰ to as high as 6‰. The minerals originate from oxidation of sulphides. The very high δ^{37}Cl values of atacamite, boleite and connellite belong to the highest ever measured and were quit unexpected. A possible explanation of these high values is as follows: During oxidation of sulphide ores, hydrogen ions (H^+) are produced according to:

$$S^{2-} + 4H_2O \leftrightarrow SO_4^{2-} + 8H^+ + 8e^- \tag{2}$$

It is supposed that due to this reaction the pH became very low and that part of the chloride escaped as gaseous hydrogen chloride with low δ^{37}Cl, so that δ^{37}Cl of the remaining solution became higher.

To test this hypothesis, the following experiment was carried out: A solution containing 0.2 M NaCl and 0.1 M H_2SO_4 was prepared. This was evaporated to dryness at 50°C and the residue was examined. Unlike the original sodium chloride, the dry residue was birefringent, and the chloride content had decreased from about 60 to 15%. Evidently, part of the chloride in the solution was lost, following the reaction:

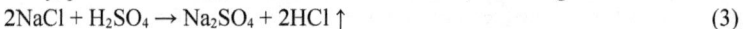

$$2NaCl + H_2SO_4 \rightarrow Na_2SO_4 + 2HCl \uparrow \tag{3}$$

The δ^{37}Cl of the evaporate was 0.52±0.05‰, thus much higher than that of the original NaCl. This shows that δ^{37}Cl may increase substantially when part of the chloride escapes as HCl gas. The result was used to make a rough estimate of the fractionation factor. A Rayleigh fractionation curve was drawn through the point were three quarters of chloride is removed with a total fractionation of 0.52. The fractionation factor ($\alpha_{gas/solution}$) based on this single measurement is about 0.99963. The Rayleigh fractionation curve in FIG. 1 predicts that a 1‰ increase in δ^{37}Cl is obtained when 93.5% of the chloride is released and a 2‰ increase is obtained when 99.58% is released. This suggest that chloride must be released almost to completion in order to get values of +6‰.

It was suggested by EGGENKAMP (1993) that the high δ^{37}Cl values in these minerals may be used as an exploration tool. For this reason the stability of atacamite was determined, assuming that stabilities for the other basic chlorides are comparable to that of atacamite. According to BARTON & BETHKE (1960) the equilibrium constant K in the

Fig. 1: Rayleigh fractionation curve indicating the $\delta^{37}Cl$ of tne residual solid as a function of the released chloride in the reaction $2NaCl+H_2SO_4>Na_2SO_4+2HCl$. $\delta^{37}Cl$ of the initial NaCl was 0‰.

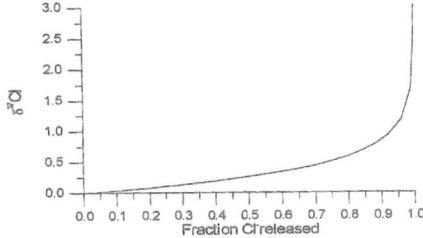

Fig. 2: Stability diagram of brochantite and atacamite as a function of SO_4^{2-} and Cl^- activity.

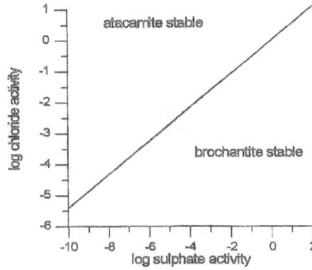

reaction brochantite ($Cu_4(OH)_6SO_4$) - atacamite ($Cu_4(OH)_6Cl_2$):

$$Cu_4(OH)_6Cl_2+SO_4 \Leftrightarrow Cu_4(OH)_6SO_4+2Cl$$

$$K=\frac{[Cl]^2}{[SO_4]} \qquad (4)$$

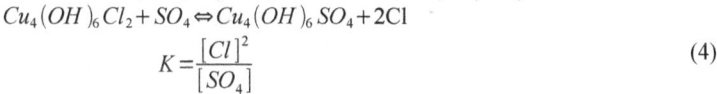

is $10^{-0.8}$. FIG. 2 shows that atacamite is only stable with respect to brochantite when:

$$[Cl]>\sqrt{0.16[SO_4]} \qquad (5)$$

Two Eh-pH diagrams are constructed for the system Cu-O-H-S-Cl. In both diagrams the total sulphur activity is 10^{-1} (this activity is chosen because it is approximately the activity of H_2S_{aq} in a solution saturated with H_2S gas at 1 atmosphere pressure at 25°C, GARRELS & CHRIST 1965). The chloride activity in FIG. 3 is 10^0 and in FIG. 4 it is 10^{-2}. Therefore, in FIG. 3 atacamite is the stable copper hydroxide mineral and in FIG. 4 brochantite is the stable copper hydroxide mineral (compare FIG. 2). In both figures total copper activities are taken as 10^{-6} (solid lines) and 10^{-4} (dotted lines). The figures are simplified in the sense that mineral phases below the line where sulphides are stable are not shown. Differences between the two diagrams occur under acid-oxidized conditions:

137

Fig. 3: Simplified Eh-pH diagram (see text) for the system Cu-O-S-Cl. Chlorine activity is 10^0, sulphur is 10^{-1} and copper activity is 10^{-4}.

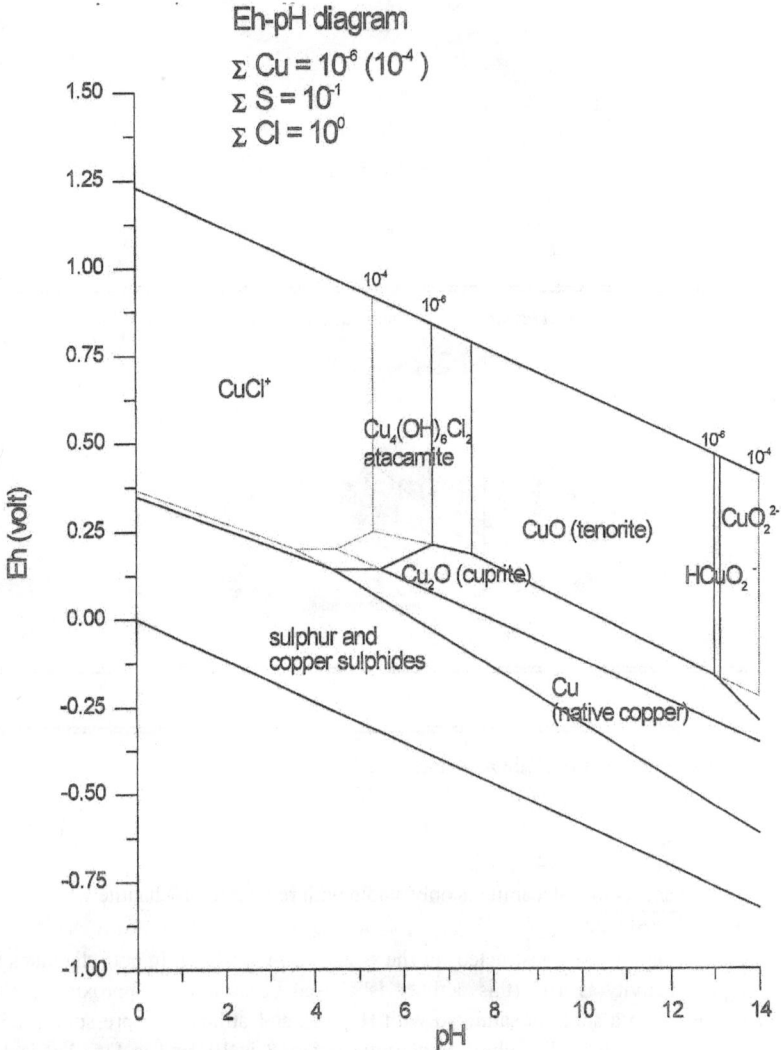

Eh-pH diagram

$\Sigma\ Cu = 10^{-6}\ (10^{-4})$
$\Sigma\ S = 10^{-1}$
$\Sigma\ Cl = 10^0$

At a total chloride activity of 10^0 the dominant species is the $CuCl^+$-ion, whereas at a total chloride activity of 10^{-2} the dominant species is the Cu^{2+}-ion. Atacamite (FIG. 3) is stable at pH-values lower than 7.55. Above this value it will lose H^+-ions and decompose to

138

Fig. 4: Simplified Eh-pH diagram (see text) for the system Cu-O-S-Cl. Chlorine activity is 10^{-2}, sulphur is 10^{-1} and copper activity is 10^{-4}.

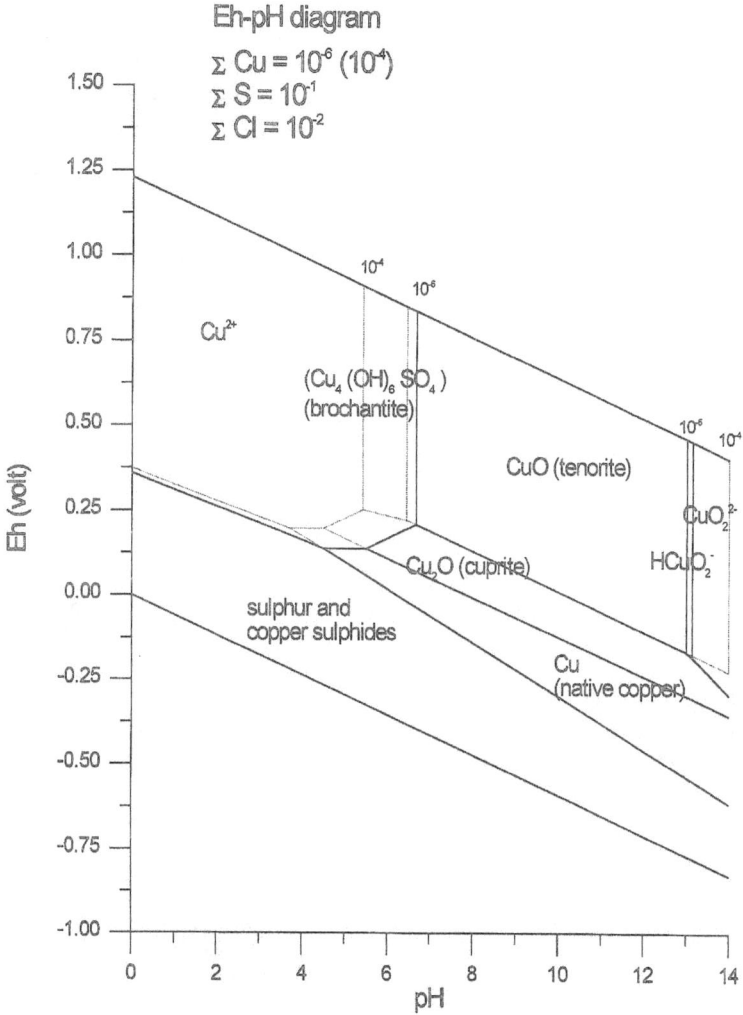

CuO (tenorite). As atacamite has a rather high solubility in acidic solutions it will dissolve at pH lower than 6.75 when the copper activity is 10^{-6}. At a copper activity of 10^{-}

[4] atacamite dissolves at pH lower than 5.42. In conclusion, atacamite is stable in neutral, oxidizing environments with high copper and chloride activities. Such environments are generally only found in deserts. In the second Eh-pH diagram (FIG. 4) brochantite is the stable copper hydroxide mineral. At a copper activity of 10^{-6}, however, brochantite is not stable; all brochantite dissolves and the Cu^{2+}-field joins the tenorite field. At a copper activity of 10^{-4} brochantite has a stability field between pH 5.42 and 6.46. The place of the brochantite field in the Eh-pH diagram is similar to that of atacamite. In conclusion, atacamite can only be formed at high chloride activities, since sulphur activities are always high in oxidizing sulphidic bodies.

Since the chlorine isotope fractionation during dissolution and precipitation of salts is not very large (chapter 10), the high $\delta^{37}Cl$ values might be used for mineral exploration. When the minerals dissolve it is expected that a fluid with relatively high $\delta^{37}Cl$ will form. This fluid can be sampled in springs and determination of $\delta^{37}Cl$ can possibly lead to the buried ore.

CONCLUSIONS

Significant variations in chlorine isotope ratios can be found in minerals with different formation histories. The various geologic settings are covered with only a limited number of unrelated samples, so that no quantitative explanations can be given. Even so, some striking systematics were found. Fumarolic minerals have negative values and basic chlorides have positive values. Values for evaporites are in agreement with earlier measured values and theoretical calculations. Magmatic minerals have $\delta^{37}Cl$ values close to SMOC, which indicates that $\delta^{37}Cl$ of the starting material had $\delta^{37}Cl$ values close to seawater. Because of the high temperatures of formation almost no fractionation appeared.

Chlorine isotopes may give important information on the genesis of chloride minerals, In particular the basic chlorides and the fumarole minerals seem to be promising subjects for further chlorine isotope research.

ACKNOWLEDGEMENTS

Most samples used were provided by the Nation Museum of Natural History in Leiden, The Netherlands. Especially we thank Prof. P.C. Zwaan, Dr C. Arps and Mr. Diederiks for their cooperation. Bas Dam provided the apatite and amphibole samples from the Bamble Area. R. Kreulen and S.O. Scholten carefully read earlier drafts of this manuscript and suggested many improvements. The mass spectrometer was partly financed by the Netherlands Organization for Scientific Research (NWO). This research is part of the AWON project "Geochemistry of chlorine isotopes" (#751.355.014).

REFERENCES

APPOLONOV Y.N., GALKIN G.A., KOSHUG D.G. & KROVOPALOV O.A. (1988) Boron mineralization in the potassic member of the Nepa deposit. *Geologiya i Geofizika* **29** 24-30
BARTON P.B. & BETHKE P.M. (1960) Thermodynamic properties of some synthetic zinc and copper

Ch. 12: $\delta^{37}Cl$ variations in minerals

minerals. *Am. J. Sci. (Bradley Vol.)* **258-A** 21-34

Braitsch O. (1962) *Entstehung und Stoffbestand der Salzlagerstätten.* Springer verlag. 232 pp.

Brögger W.C. & Reusch H.H. (1875) Vorkommen des Apatit in Norwegen. *Zs. Deuts. Geol. Ges.* **27** 646-702

Cuitiño L. (1986) Mineralogia y genesis del yacemento de lapislazuli Flor de los Andes, Coquimbo, Norte de Chile. *Rev. Geol. Chile* **27** 55-67

Curie I. (1921) Sur le poids atomique du chlore dans quelques mineraux. *Comp. Rendus Seances* **172** 1025

Dam B.P. (1994) Ph.D. Thesis. *in prep.*

Deer W.A., Howie R.A. & Zussman J. (1966) *Rock-forming minerals. Vol. 4.* Green & Co, London. 435 pp.

Dorenfelt M. (1923) Relative determination of the atomic weight of chlorine in Bamle apatite. *J. Amer. Chem. Soc.* **45** 1577-1579

Eggenkamp H.G.M. (1993) Can chlorine stable isotope ratio variations be used in mineral exploration? *Abstr. vol. 16th IGES, Beijing.* pp. 37-38

Förtsch E. (1970) Untersuchungen an Mineralien der Pyromorphit-Gruppe. *N. Jb. Miner. Abh.* **113** 219-250

Garrels R.M. & Christ C.L. (1965) *Solutions, minerals, and equilibria.* Harper & Row. New York. 450 pp.

Gleditsch E. & Samdahl B. (1922) Sur le poids atomique du chlore dans un mineral ancien, l'apatite de Bamle. *Comp. Rendus Seances* **174** 746

Heide K., Franke H. & Brückner H.-P. (1980) Vorkommen und Eigenschaften von Boracit in den Zechsteinsalzlagerstatten der DDR. *Chem. Erde* **39** 201-232

Hintze C. (1915) *Handbuch der Mineralogie.* Verlag von Veit & Comp. Leipzig.

Hodenberg R.v., Fischbeck R. & Kühn R. (1987) Beitrag zur Kenntnis der Salzminerale, Salzgesteine und Salzlagerstätten, insbesondere im deutschen Zechstein. *Aufschluss* **38** 109-125

Hoering T.C. & Parker P.L. (1961) The geochemistry of the stable isotopes of chlorine. *Geochim. Cosmochim. Acta* **23** 186-199

Hölzel A.R. (1989) *Systematics of minerals.* 1st. Edition. 584 pp.

Humphreys D.A., Thomas J.H. & Williams P.A. (1980) The chemical stability of mendipite, diaboleïte, chloroxiphite, and cumengéite, and their relationships to other secundary lead(II) minerals. *Min. Mag.* **43** 901-904

Kaufmann R.S. (1989) Equilibrium exchanche models for chlorine stable isotope fractionation in high temperature environments. *Proc. WRI.* **6** 365-368

Krivdick S.G. & Tkachuk V.I. (1988) Eudialite-bearing agpaitic phonolites and dike nepheline syenites in the October intrusion, Ukrainian shield. *Geokhimiya* **1988** 1133-1139

Krupp R., Nottes G. & Heitke U. (1985) Zur Kenntnis einiger natürlich vorkommender Quecksilber-Halogen-Verbindungen, besonders des Jodquecksilbers, im Moschellandsberg/Pfalz. *Aufschluss* **36** 73-80

Lacroix M.A. (1907) Les minéraux des fumarolles de l'eruption du Vésuve en avril 1906. *Bull. Soc. Min. Fr.* **30** 219-266

Morton R.D. & Catanzaro E.J. (1964) Stable chlorine isotope abundances in apatites from Ødegårdens Verk, Norway. *Norsk Geol. Tiddsk.* **44** 307-313

Owen H.R. & Schaeffer O.A. (1954) The isotope abundances of chlorine from various sources. *J. Amer. Chem. Soc.* **77** 898-899

Rodriguez E. (1980) Condiciones de formacion de algunos minerales oxidados de cobre. *Rev. Geol. Chile* **9** 57-61

Smirnov S.S. (1954) *Die Oxydationszone Sulphidischer Lagerstätten.* Akademie Verlag, Berlin

Vertacnik W. (1983) Salmiak von der Abraumhalde der ehemaligen Zeche Altstaden, Rheinland.

Aufschluss **34** 177-180

CHAPTER 13

Chlorine Stable Isotopes in Carbonatites

H.G.M. Eggenkamp[1], A.F. Koster van Groos[2] and R. Kreulen[1]

ABSTRACT-- $\delta^{37}Cl$, $\delta^{18}O$ and $\delta^{13}C$ were measured in eight different carbonatites. Correlations between $\delta^{37}Cl$, $\delta^{18}O$ and $\delta^{13}C$ are suggested, indicating possible alteration with meteoric water or Rayleigh fractionation in secondary carbonatites. Primary carbonatites have negative $\delta^{37}Cl$ values. Probably, these samples are mantle material, which might indicate that the mantle is impoverished in ^{37}Cl. Since surface chloride mainly is derived from degassed mantle material, it is supposed that HCl escaping from the mantle is enriched in ^{37}Cl. This is in agreement with results obtained from volcanic water and gas samples (chapter 8). Since the surface reservoir in this way is enriched in ^{37}Cl, the residual chlorine in the mantle must have lower $\delta^{37}Cl$ values than this surface reservoir.

INTRODUCTION

Carbonatites are igneous rocks containing at least 50% carbonate minerals and at most 10% SiO_2 (WOOLLEY & KEMPE 1989). The carbonate minerals usually are calcite, dolomite, siderite, or, rarely, sodium carbonate. Because carbonatites originate from the mantle, they may provide information on the chlorine isotope composition of the mantle. In this study we present the first chlorine isotopic compositions of calcium, iron, and sodium-rich carbonatites. Oxygen and carbon isotopes were measured as well for comparison.

-Chlorine in carbonatites

Chloride contents in carbonatites range between 0 and 0.45% (WOOLLEY & KEMPE 1989). Chloride content appears to decrease from iron-poor (avg. 0.08% Cl⁻) to iron-rich (avg. 0.02% Cl⁻) carbonatites (WOOLLEY & KEMPE 1989), although only few carbonatites were analyzed for chloride. In African carbonatites 12 to 1023 ppm Cl was reported (DAWSON & FUGE 1980), but their iron content was not given. In the unusual, extrusive natrocarbonatite of Oldoinyo Lengai, the chloride content is very high, 5.50% (KELLER & KRAFFT 1990). In carbonatites, chlorine occurs mainly in apatites, micas and amphiboles. Very little is known about the behaviour of chlorine in carbonatite melts, especially concerning the solubility and partition coefficients between the solid and liquid phase (GITTINS 1989).

1Department of Geochemistry, Utrecht University, P.O.Box 80.021, 3508 TA Utrecht, The Netherlands
2Department of Geological Sciences, The University of Illinois at Chicago, Box 4348, Chicago, Illinois 60680, USA

-$\delta^{18}O$ and $\delta^{13}C$ in carbonatites

D EINES (1989) reviewed literature data on $\delta^{18}O$ and $\delta^{13}C$ variations in carbonatites. $\delta^{18}O$ varies from about +4 to +27‰, but primary unaltered carbonatites range between +6 and +9‰. Thus, average $\delta^{18}O$ of carbonatites is slightly higher than mantle $\delta^{18}O$, which ranges between +5 and +6‰ as found in mafic and ultramafic xenoliths and unaltered basalts. Equilibrium $\delta^{18}O$ fractionation between carbonates and silicates from the mantle is about 2‰ at 1000 °C. Therefore D EINES (1989) concluded that oxygen in carbonatites originates from the mantle.

$\delta^{13}C$ variations are more restricted than the $\delta^{18}O$ values. Most data are between -2 and -8‰ vs. PDB, and no extreme values are found. Carbon is a minor element in the mantle and probably not homogeneously distributed. Large variations in $\delta^{13}C$ are found in stony meteorites, mafic and ultramafic xenoliths, unaltered basalts, and diamonds. The most reliable $\delta^{13}C$ data are from diamonds. Carbonatites $\delta^{13}C$ values agree with the average $\delta^{13}C$ in diamonds, but the range in diamonds is much larger (D EINES 1989).

Generally, no correlation was observed between $\delta^{18}O$ and $\delta^{13}C$ in carbonatites. However, a weak correlation may exist at the $\delta^{18}O$ range between 5.5 and 14.5‰, where $\delta^{13}C$ increases from an average value of about -5.5 to -2.5‰ (D EINES 1989). "Primary", mantle derived carbonatites have coexisting calcites and dolomites with very similar isotope values. Some carbonatites have high $\delta^{18}O$ values, which strongly suggest interaction with groundwater or recrystallization (S HEPPARD & D AWSON 1973). Thus, these carbonatites show evidence of alteration.

MATERIAL

Eight samples of powdered carbonatites were used in this study. Sample MFCA is a miocene sövite from the Orberg/Schelingen locality from the Kaiserstuhl, S.W.Germany. It was kindly provided by Prof. J. Keller. A sövite is, a calcite rich carbonatite, according to the definition by B RÖGGER (1921).

The other seven samples are from Africa. One is a natrocarbonatite from Oldoinyo Lengai (OL112, K ELLER & K RAFFT 1990, collected in 1988 from the lava lake, kindly provided by Prof. J. Keller). The other samples were provided by dr M.J. Le Bas. Two samples are from South Africa, 86CN8 (a clinohumite-phlogopite sövite from the Nooitgedacht complex, Transvaal) and 85CK64 (a magnetite bearing sövite from the Kruidfontein complex, Transvaal). These samples are of proterozoic age (e.g. C LARKE & LE B AS 1990). Sample HF509 is from the north side of the Homa mountain complex, W. Kenia (F LEGG *et al.* 1977). This pyrochlore-sodic amphibole sövite is of Tertiary age. The last three samples are from the North Ruri complex, Ruri Hills, near Homa bay, W. Kenya. They also are of tertiary age (D IXON & C OLLINS 1977). These are N428, a biotite-aegirine sövite, N340, an alkvite containing minor magnetite, and N423, a ferrocarbonatite which contains some biotite. An alkvite is a dolomitic carbonatite according to the definition by B RÖGGER (1921).

METHODS

Most carbonatites contain little chloride, with the exception of the natro-carbonatites. Chlorine was extracted by the following method. 5 gram of carbonatite (except Oldoinyo Lengai) was mixed with 100 ml distilled water and stirred at a temperature of 50 °C. To this mixture, concentrated HNO_3 was added (drop by drop), until the solution stopped to sparkle (pH is then as low as 0.5). Next, the solution was filtered. To the filtrate was added 6 gram KNO_3, 2.06 gram citric acid and 0.07 gram $Na_2HPO_4.2H_2O$ (chapter 2). Only the chloride in the carbonate and the apatite, (and not in the silicates) is liberated with this method. From this solution, AgCl was precipitated. After reaction of AgCl with CH_3I, the resulting CH_3Cl was measured on the mass spectrometer (e.g. KAUFMANN 1984). The chloride content of the carbonatites was calculated from the yield of CH_3Cl.

$\delta^{18}O$ and $\delta^{13}C$ are determined on CO_2 which was produced by reacting the carbonatite with phosphoric acid at 25 °C (McCREA 1950). Carbonates were reacted for 4 hours, so $\delta^{13}C$ and $\delta^{18}O$ of carbonates less reactive than calcite are not measured.

Table 1: Results of chloride concentrations and isotope measurements on carbonatites.

Code	Cl⁻ (ppm)	$\delta^{37}Cl$	$\delta^{13}C$	$\delta^{18}O$
MFCA	115	-0.01±0.08	-5.82±0.02	7.52±0.12
HF509	91	-0.22±0.08	-4.48±0.03	8.84±0.19
86CN8	64	-0.36±0.18	-4.73±0.03	7.55±0.04
85CK64	109	0.61	-2.66±0.03	13.39±0.14
N428	21	-0.84	-5.06±0.03	10.19±0.16
N340	15	-0.38±0.36	-4.04±0.01	18.84±0.25
N423	21	-0.10±0.58	-2.62±0.01	25.11±0.05
OL112	41597	+0.13±0.05	-7.16±0.01	5.70±0.10

RESULTS

In **table 1** the extracted chloride concentration and the isotope data are listed. The range in chloride content was large. Extractable chloride content in all but the natrocarbonatite sample between 15 and 115 ppm. The total chloride content is generally much higher, for example, in MFCA it is about 700 ppm (J. Keller, pers. comm.). The difference between total and extractable chloride is caused by the larger part of the chloride incorporated in minerals that are not soluble in HNO_3, such as micas and amphiboles. The reason we choose the HNO_3 extraction method is that in order to extract all chloride from the sample, the 5 gram of sample has to be dissolved in 50 gram of molten NaOH. This would result in a large quantity of fluid with considerable risk of contamination. For OL112, the difference between total chloride (5.5%, KELLER & KRAFFT 1990) and extractable chloride is less. In this sample Cl^-, together with SO_4^{2-}, PO_4^{3-} and F^-, substitutes for CO_3^{2-} in nyerereite $(Na_{0.82}K_{0.19})_2(Ca,Sr,Ba)_{0.975}(CO_3)_2$ and gregoryite $Na_{1.74}K_{0.1}(Ca,Sr,Ba)_{0.16}CO_3$ (DAWSON 1989, KELLER & KRAFFT 1990).With the exception of OL112, two groups can be distinguished, one with low concentrations (the three Ruri samples) and one with higher concentrations (the other samples). $\delta^{37}Cl$ values range from -0.84 to +0.61‰. Unfortunately, the most positive and the most negative sample (85CK64 and N428) could only be measured once because there was not enough sample material. The standard deviation of the measurements of the samples N340 and N423 is too large, due to the very low chloride concentrations. Remeasuring was impossible because of insufficient material.

$\delta^{18}O$ ranges from +5.70 to +25.11‰ and $\delta^{13}C$ from -7.16 to -2.62‰. Thus, the measured samples cover the complete range of values found in previous studies on carbonatites (DEINES 1989).

DISCUSSION

Very few literature data exist on the chloride geochemistry of carbonatites. In general (except for Oldoinyo Lengai) the chloride concentrations are low. However, since the chloride concentrations were not measured in most cases, it is not known whether regional variations occur. WOOLLEY & KEMPE (1989) reviewed analyses of 230 carbonatites, but only 12 of them include chloride analyses (i.e. only 5.2% of the total).

There is a large database of $\delta^{18}O$ and $\delta^{13}C$ values (see DEINES 1989). Carbonatites with $\delta^{18}O$ between +6 and +9‰ are considered primary mantle material. $\delta^{13}C$ of primary carbonatites is supposed to be between -2 and -8‰. Based on the $\delta^{18}O$ values, our samples MFCA, 86CN8 and HF509 represent primary carbonatites. The other samples have higher or lower $\delta^{18}O$ values and secondary processes, such as interaction with groundwater, probably occurred.

The three Ruri samples (N428, N340 and N423) are from the same carbonatite. In these samples all three measured isotopes are positively correlated (see FIG. 1). The $\delta^{18}O$ value of N423 (+25.11‰) is one of the highest values ever measured in a carbonatite. The extremely high $\delta^{18}O$ value is difficult to explain by hydrothermal alteration with meteoric water. Meteoric water and this particular carbonatite would be in isotopic equilibrium at ambient temperature. At higher temperatures the $\delta^{18}O$ of the carbonatite will become lower. $\delta^{18}O$ of this carbonatite is so high that it even can be supposed to have been originally a marble.

Since no other $\delta^{37}Cl$ measurements on mantle material are known, no comparisons similar to those for $\delta^{18}O$ and $\delta^{13}C$ can be made. Although $\delta^{18}O$ of sample N428 was outside the range for primary carbonatites, it could indicate very low $\delta^{37}Cl$ values in primary carbonatites. If the correlation for the isotopes is extrapolated to lower $\delta^{18}O$ values, $\delta^{37}Cl$ for primary carbonatites in this system would be lower than -1‰. This

146

Fig. 1: Results of isotope measurements of the carbonatites, including correlations in the Ruri samples.

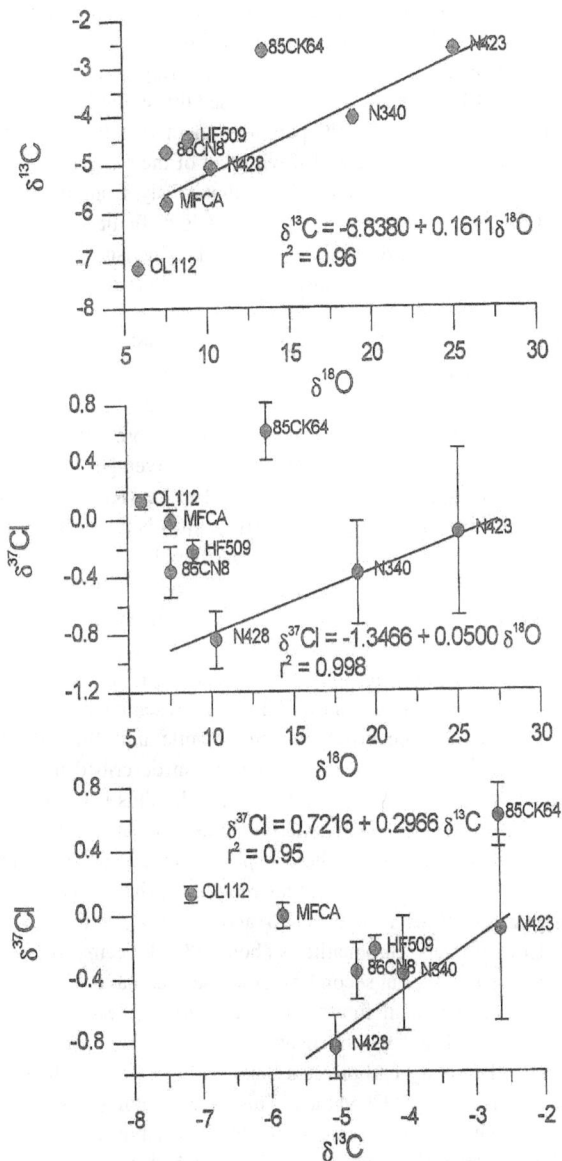

suggests that $\delta^{37}Cl$ of mantle material can be very negative.

A positive correlation between $\delta^{37}Cl$ and both $\delta^{18}O$ and $\delta^{13}C$ is also suggested by the African samples other than the Ruri samples (86CN8, HF509 and 85CK64). These samples are less iron rich and $\delta^{18}O$ is near the limit of the primary carbonatite field or a few per mil higher. If the $\delta^{37}Cl$ variations are caused by the same process that produced the correlation between $\delta^{18}O$ and $\delta^{13}C$, it is possible that Rayleigh fractionation took place as proposed by DEINES (1989). DEINES (1989) supposed this could lead to $\delta^{18}O$ values up to +15‰ and that the slope in a $\delta^{18}O$-$\delta^{13}C$-plot was about 0.4. It is not unlikely that the fractionation of chlorine isotopes is coupled with that of the oxygen and carbon isotopes. This implies that two phases are removed simultaneously from the same reservoir, a carbonaceous and a siliceous one. At the moment it is difficult to tell whether this is a viable explanation, because fractionation factors for chlorine isotope partitioning over the relevant phases are not known. Since chlorine is only a minor element in all phases this is important to know.

The natrocarbonatite of Oldoinyo Lengai most probably originated by immiscible separation from a carbonated olivine nephelinite magma deep in the mantle (LE BAS 1987, TWYMAN & GITTINS 1987, DAWSON 1989). Its chloride concentration is extremely high. It is for the larger part present in sylvite (which is present in the lava for about 10%), which is a primary mineral here. In gregoryite Cl⁻ averages 0.5% and in nyerereite it averages 0.26%. The origin of this chloride is not clear yet (KELLER & KRAFFT 1990).

From this it follows that $\delta^{37}Cl$ of primary carbonatites is not very constant and varies between -1 and 0‰. As these carbonatites are considered primary mantle material, this can be an indication that $\delta^{37}Cl$ of the mantle also is negative. Unfortunately carbonatites represent no unaltered mantle material. Carbonatite magmas are very fractionated magmas. It is not impossible that chlorine isotopes are also fractionated during the formation of the carbonatite magma. In this case the chlorine isotope composition is not representative for the mantle. Some causes for not representative $\delta^{37}Cl$ values could be for example 1) fractionation occurred between mantle and fluid during carbonatite extraction, 2) diffusion fractionation comparable to that described in chapter 10, 3) the carbonatites represent a different system than the mantle, thus is not in equilibrium with the mantle. carbonatites are very fractionated magmas. For all these explanations much more information is needed. Although these explanations can not be rejected here, it is suggested here that the measured $\delta^{37}Cl$ values represent primary mantle values. $\delta^{37}Cl$ in measured primary carbonatites is between 0 and -0.36‰. In the Ruri samples it is supposed that $\delta^{37}Cl$ of primary carbonatite is about -1‰. It seems that the mantle is not very homogeneous in chloride, but secondary processes can have their influence. For this reason it is recommended that in future chlorine isotope research of carbonatites also $\delta^{37}Cl$ in the less affected silicates will be measured.

Since all supposed primary carbonatites have negative $\delta^{37}Cl$ values, it is concluded that the mantle has negative $\delta^{37}Cl$ values. This is in striking agreement with results obtained in volcanic water samples (chapter 8). In that chapter it is tentatively proposed that chloride outgassed from the mantle as HCl is enriched in ^{37}Cl, and thus that $\delta^{37}Cl$ in the residual rock is negative. It is assumed that surface chloride originated from outgassing of mantle chloride (e.g. Schilling *et al.* 1978), As it is proposed that this gas could be enriched in ^{37}Cl this is in agreement with the measured $\delta^{37}Cl$ values in carbonatites relative to seawater. Of course this is based on only eight measurements thus it is highly recommended to measure more material. Not only carbonatites, but also other mantle material has to be measured. To determine fractionation factors also experimental work has to be done.

CONCLUSIONS

This study presents the first chlorine isotope measurements ever on carbonatites. Significant variations in $\delta^{37}Cl$ were observed, but since the samples came from different geological settings it is difficult to interpret the results in terms of fractionating processes. A correlation between chlorine, oxygen and carbon isotopes is suggested. If fractionation of chlorine isotopes is caused by the same processes as the fractionation of oxygen and carbon isotopes, then exchange with the country-rocks or Reyleigh fractionation may be important. The data suggest that $\delta^{37}Cl$ in primary carbonatites can be between -1 and 0‰. If these data represent mantle values, which is not proven, this is in agreement with data of volcanic water samples from Indonesia which indicate that chloride escaped from the mantle is enriched in ^{37}Cl, resulting in negative $\delta^{37}Cl$ values in the mantle.

The results invite more research. It is clear that further research must be combined with an effort to understand the chloride geochemistry of carbonatites, since almost nothing is known on this subject. It also is important to obtain good fractionation knowledge of the HCl degassing system. It will be important to measure both the HNO_3 extractable and the total chloride.

ACKNOWLEDGEMENTS

Samples MFCA and OL112 were provided by Prof. J. Keller. All other samples were provided by Dr M.J.J. Le Bas. Dr S.O. Scholten carefully read an earlier draft of this manuscript. The mass spectrometer is partly financed by the Netherlands Organization of Scientific Research. This research is part of the AWON-project "Geochemistry of chlorine isotopes" (№755.351.014).

REFERENCES

Brögger W.C. (1921) Die Eruptivgesteine des Kristianagebietes. IV. Das Fengebied in Telemark, Norwegen. *Videnskapsselskapets Skrifter. I. Mat-Naturv. Klasse* **1920** No. 9

Clarke L.B. & Le Bas M.J. (1990) Magma mixing and metasomatic reaction in silicate-carbonate liquids at the Kruidfontein carbonatitic volcanic complex, Transvaal. *Miner. Mag.* **54** 45-56

Dawson J.B. (1989) Sodium carbonatite extrusions from Oldoinyo Lengai, Tanzania: Implications for carbonatite complex genesis. in *Carbonatites, genesis and evolution.* ed. K. Bell 255-277

Dawson J.B. & Fuge R. (1980) Halogen content of some african primary carbonatites. *Lithos* **13** 139-143

Deines P. (1989) Stable isotope variations in carbonatites. in *Carbonatites, genesis and evolution.* ed. K. Bell 301-359

Dixon J.A. & Collins B.A. (1977) The carbonatite complex of North and south Ruri. in *Carbonatite-Nephelinite Volcanism* ed. M.J. Le Bas 169-198

FLAGG A.M., CLARKE M.C.G., SUTHERLAND D.S. & LE BAS M.J. (1977) Homa Mountain II: The main carbonatite. **in** *Carbonatite-Nephelinite Volcanism* ed. M.J. LE BAS 222-232

GITTINS J. (1989) The origin and evolution of carbonatite magmas. **in** *Carbonatites, genesis and evolution.* ed. K. BELL 580-600

KAUFMANN R.S. (1984) *Chlorine in groundwater. Stable isotope distribution.* Ph.D. thesis. Arizona, 137 pp.

KELLER J. & KRAFFT M. (1990) Effusive natrocarbonatite activity of Oldoinyo Lengai, June 1988. *Bull. Volanol.* **52** 629-645

LE BAS M.J. (1987) Nephelinites and carbonatites. In: *Alkaline igneous rocks.* Edt. J.G. FITTON & B.G.J. UPTON. *Geol. Soc. Spec. Publ.* **30** 53-83

McCREA J.M. (1950) The isotope chemistry of carbonates and a paleotemperature scale. *J. Chem. Phys.* **18** 849-857

SCHILLING J.-G., UNNI C.K. & BENNING M.L. (1978) Origin of chlorine and bromine in the oceans. *Nature* **273** 631-636

SHEPPARD S.M.F. & DAWSON J.B. (1973) $^{13}C/^{12}C$, $^{18}O/^{16}O$ and D/H isotope variations in "primary" igneous carbonatites. *Fortsch. Mineral.* **50** 128-129

TWYMAN J.D. & GITTINS J. (1987) Alkalic carbonatite magmas: Parental or derivative? In: *Alkaline igneous rocks.* Edt. J.G. FITTON & B.G.J. UPTON. *Geol. Soc. Spec. Publ.* **30** 85-94

WOOLLEY A.R. & KEMPE D.R.C. (1989) Carbonatites: nomenclature, average chemical compositions, and element distribution. **in** *Carbonatites, genesis and evolution.* ed. K. BELL 1-14

Curriculum Vitæ

Hans Eggenkamp was born on the 22nd October 1963 in Laren (NH), The Netherlands. After secondary school (Laar en Berg, Laren), he started to study Geology at Utrecht University, where his Masters degree in Geochemistry was obtained in 1988. Between 1988 and 1994 the work described in this thesis was performed. Between 1994 and 1998 he was appointed as Post-Doctoral Research Fellow at the University of Reading (UK), from 1998 until 2000 as Lecturer at Utrecht University and from 2000 until 2008 as Geochemist at Isolab BV (Neerijnen, The Netherlands). Since 2008 he is appointed as Investigador Auxiliar at the Instituto Superior Técnico (Universidade Técnica de Lisboa, Portugal). During this period he continued to work on the Geochemistry of Chlorine Isotopes.

www.ingramcontent.com/pod-product-compliance
Lightning Source LLC
Chambersburg PA
CBHW071553200326
41519CB00021BB/6733